Lecture Notes in Mathematics

Edited by A. Dold and B. Eckmann

1265

Walter Van Assche

Asymptotics for Orthogonal Polynomials

Springer-Verlag
Berlin Heidelberg New York London Paris Tokyo

Author

Walter Van Assche
Senior Research Assistant of the Belgian National Fund for Scientific Research)
Departement Wiskunde, Katholieke Universiteit Leuven
Celestijnenlaan 200 B, 3030 Leuven, Belgium

Mathematics Subject Classification (1980): Primary: 42C05
Secondary: 33A65; 39A10; 40A15; 40A30; 41A55; 41A60; 60J80

ISBN 3-540-18023-0 Springer-Verlag Berlin Heidelberg New York
ISBN 0-387-18023-0 Springer-Verlag New York Berlin Heidelberg

Printed in Germany

Printing and binding: Druckhaus Beltz, Hemsbach/Bergstr.
2146/3140-543210

PREFACE

In recent years there has been an increased interest in the theory of orthogonal polynomials but the number of textbooks treating orthogonal polynomials is rather limited. Even at present the best reference is Szegö's book [175] which was first published in 1939. Freud's book [61] is also highly recommended. The more recent introduction by Chihara [39] does not include asymptotic results and the monographs by Geronimus [75] and Nevai [138] are rather technical and treat a very specific part of the theory of orthogonal polynomials on [-1,1]. This monograph concentrates on the asymptotic theory of general orthogonal polynomials on the real line. Most of the theorems have been proved, for some of them only a sketch of the proof is given and tedious proofs out of the scope of this monograph have been omitted.

I would like to express my very cordial thanks to all those who, in some way or another, have contributed to this monograph. Many thanks in particular to Jef Teugels who made me appreciate the mathematical analysis of orthogonal polynomials. Some theorems in this monograph are the result of working with other mathematicians. I am very grateful to Makoto Maejima (Chapter 3), Jeffrey Geronimo (Chapter 2, Section 4.4) and Guido Fano (Section 5.1) for a fruitful collaboration. Finally I would like to thank Daniel Bessis, Pierre Moussa, Giorgio Turchetti, Paul Nevai, Doron Lubinsky, Ed Saff, Alphonse Magnus and many others for various interesting discussions and Bea Peeters for an excellent job of typing the manuscript.

Walter Van Assche, March 1987.

TABLE OF CONTENTS

INTRODUCTION

0.1. Definitions and examples

Let μ be a positive probability measure on the real line with distribution function $\mu(t) = \mu((-\infty,t])$. Suppose that all the *moments*

$$m_n = \int_{-\infty}^{\infty} x^n d\mu(x) \tag{0.1.1}$$

are finite and that the *support* of the measure μ

$$\operatorname{supp}(\mu) = \{x \in \mathbb{R} : \forall\varepsilon > 0 \quad \mu((x-\varepsilon,x+\varepsilon)) > 0\}$$

is an infinite set. Then there exists a sequence of polynomials $\{p_n(x) : n = 0,1,2,\dots\}$ such that

$$\int_{-\infty}^{\infty} p_n(x)p_m(x)d\mu(x) = \delta_{m,n} \qquad m,n \geq 0$$

(0.1.2)

$$p_n(x) = p_n(x;\mu) = k_n x^n + \dots \;, \qquad k_n > 0 \;.$$

This sequence consists of *orthogonal polynomials* with *spectral measure* μ (or *orthogonality measure*) (Szegö [175], p. 23). The definition (0.1.2) actually implies that the polynomials are normed so that one should speak of "orthonormal polynomials". We denote the *monic polynomials* by

$$\hat{p}_n(x) = k_n^{-1}\, p_n(x) \;. \tag{0.1.3}$$

Since μ is a probability measure, it follows that

$$p_0(x) = \hat{p}_0(x) = 1.$$

It is possible to extend the notion of orthogonal polynomials by using a measure on some curve in the complex plane, but in this monograph we will always use orthogonality on the real line (the only exception is § 1.4).

The measure μ can always be decomposed as a linear combination of three different types of measures, $\mu = \mu_{ac} + \mu_d + \mu_s$, μ_{ac} is an absolutely continuous measure (with respect to Lebesgue measure), μ_d is an atomic measure with mass on a discrete set which is at most denumerable and μ_s is a singular measure (with respect to Lebesgue measure) with a continuous distribution function. If the spectral measure is absolutely continuous, then there exists a (Radon-Nikodym) derivative w such that

$$\int_{-\infty}^{\infty} p_n(x)p_m(x)w(x)dx = \delta_{m,n} \qquad m,n \geq 0 .$$

The function w is then the *weight function* for the orthogonal polynomials.

Example 1 (Szegö [175], Chapter 4).

Consider the weight function ($\alpha,\beta > -1$)

$$\text{(0.1.4)} \qquad w(x) = \begin{cases} 2^{-\alpha-\beta-1} \dfrac{\Gamma(\alpha+\beta+2)}{\Gamma(\alpha+1)\Gamma(\beta+1)} (1-x)^{\alpha}(1+x)^{\beta} & -1 < x < 1 \\ 0 & \text{elsewhere} \end{cases}$$

(this is the density of a beta-distribution on [-1,1]), then

$$\text{(0.1.5)} \qquad p_n(x) = \left\{\frac{2n+\alpha+\beta+1}{n+\alpha+\beta+1} \frac{n!(\alpha+\beta+2)_n}{(\alpha+1)_n\ (\beta+1)_n}\right\}^{1/2} P_n^{(\alpha,\beta)}(x)$$

where $\{P_n^{(\alpha,\beta)}(x) : n = 0,1,2,\ldots\}$ are *Jacobi polynomials* with parameters α and β. We have used the Pochhammer notation

$$(a)_n = a(a+1)\ \ldots\ (a+n-1) = \frac{\Gamma(a+n)}{\Gamma(a)} .$$

The Jacobi polynomials are explicitely given by

$$\text{(0.1.6)} \qquad P_n^{(\alpha,\beta)}(x) = 2^{-n} \sum_{j=0}^{n} \binom{n+\alpha}{j} \binom{n+\beta}{n-j}(x-1)^{n-j}(x+1)^j$$

$$= \frac{\Gamma(\alpha+n+1)}{n!\Gamma(\alpha+\beta+n+1)} \sum_{j=0}^{n} \binom{n}{j} \frac{\Gamma(\alpha+\beta+n+j+1)}{\Gamma(\alpha+j+1)} \left(\frac{x-1}{2}\right)^j .$$

They satisfy the differential equation

$$(1-x^2)y'' + [\beta-\alpha-(\alpha+\beta+2)x]y' + n(n+\alpha+\beta+1)y = 0$$

and can be found from *Rodrigues' formula*

$$(1-x)^{\alpha}(1+x)^{\beta}\, P_n^{(\alpha,\beta)}(x) = \frac{(-1)^n}{2^n n!}\frac{d^n}{dx^n}\{(1-x)^{n+\alpha}(1+x)^{n+\beta}\} .$$

Special cases include the *Legendre polynomials*

$$P_n(x) = P_n^{(0,0)}(x) ,$$

the *Chebyshev polynomials of the first kind*

$$T_n(x) = \frac{2.4\ldots 2n}{1.3\ldots(2n-1)}\, P_n^{(-1/2,-1/2)}(x) ,$$

and the *Chebyshev polynomials of the second kind*

$$U_n(x) = \frac{1}{2}\,\frac{2.4\ldots(2n+2)}{1.3\ldots(2n+1)}\, P_n^{(1/2,1/2)}(x) .$$

If we set $x = \cos t$ then

$$T_n(x) = \cos(nt) \quad ; \quad U_n(x) = \frac{\sin((n+1)t)}{\sin t} .$$

Jacobi polynomials with $\alpha = \beta$ are called *ultraspherical* or *Gegenbauer polynomials*.

Example 2 (Szegö [175], Chapter 5).

The orthogonal polynomials for the weight function ($\alpha > -1$)

$$(0.1.7) \qquad w(x) = \begin{cases} \frac{1}{\Gamma(\alpha+1)}\, x^{\alpha} e^{-x} & x > 0 \\ 0 & x < 0 \end{cases}$$

are given by

$$(0.1.8) \qquad p_n(x) = (-1)^n \binom{n+\alpha}{n}^{-1/2} L_n^{(\alpha)}(x)$$

where

$$(0.1.9) \qquad L_n^{(\alpha)}(x) = \sum_{j=0}^{n} \binom{n+\alpha}{n-j} \frac{(-x)^j}{j!} .$$

The polynomials in (0.1.9) are the *Laguerre polynomials* when $\alpha = 0$; for arbitrary $\alpha > -1$ these polynomials are called *generalized Laguerre polynomials* or *Sonin-Laguerre polynomials*. They satisfy the differential equation

$$xy'' + (\alpha+1-x)y' + ny = 0$$

and can be obtained through Rodrigues' formula

$$e^{-x}x^{\alpha}L_n^{(\alpha)}(x) = \frac{1}{n!}\frac{d^n}{dx^n}(e^{-x}x^{n+\alpha}) .$$

Notice that (0.1.7) is the density of a gamma distribution on $[0,\infty)$.

Example 3 (Szegö [175], Chapter 5).

Consider the density of a normal (or Gaussian) distribution

(0.1.10) $$w(x) = \frac{1}{\sqrt{\pi}} e^{-x^2} \qquad x \in \mathbb{R}$$

then the orthogonal polynomials are

(0.1.11) $$p_n(x) = (2^n n!)^{-1/2} H_n(x)$$

where $\{H_n(x) : n = 0,1,2,\dots\}$ are the *Hermite polynomials*. Somewhat more general is the weight function ($\alpha > -1/2$)

(0.1.12) $$w(x) = \frac{1}{\Gamma(\alpha+\frac{1}{2})} |x|^{2\alpha}e^{-x^2}$$

with

(0.1.13) $$p_n(x) = 2^{-n}\{[\tfrac{n}{2}]!\,(\alpha + 1/2)_{[\frac{n+1}{2}]}\}^{-1/2} H_n^{(\alpha)}(x) .$$

The polynomials $\{H_n^{(\alpha)}(x) : n = 0,1,2,\dots\}$ are *generalized Hermite polynomials* or *Markov-Sonin polynomials*. They satisfy the differential equation

$$xy'' + 2(\alpha - x^2)y' + (2xn - \theta_n x^{-1})y = 0$$

with $\theta_{2m} = 0$ and $\theta_{2m+1} = 2\alpha$. There exists a Rodrigues' formula which is (for $\alpha = 0$)

$$e^{-x^2}H_n(x) = (-1)^n \frac{d^n}{dx^n}(e^{-x^2}) .$$

There is a simple relation with the (generalized) Laguerre polynomials :

(0.1.14) $$H_{2n}^{(\alpha)}(x) = (-1)^n\, 2^{2n}\, n!\, L_n^{(\alpha-1/2)}(x^2)$$

(0.1.14') $H^{(\alpha)}_{2n+1}(x) = (-1)^n\, 2^{2n+1}\, n! \times L^{(\alpha+1/2)}_n(x^2)$.

The polynomials of Jacobi, Laguerre and Hermite together are the *classical orthogonal polynomials*. They can be characterized as being the only ones that satisfy a homogeneous linear differential equation of the second order. This class also consists of the only orthogonal polynomials for which the derivatives are again orthogonal polynomials. A formula of Rodrigues type is also possible only within this class.

There are other important sequences of orthogonal polynomials. We might call those semi-classical orthogonal polynomials. They are obtained by allowing a discrete version of the notion of derivative (ordinary difference or q-difference). Some examples are

Example 4 (Chihara [39], p. 170).

Suppose the spectral measure is discrete and supported on the positive integers with jumps $(a > 0)$

(0.1.15) $\mu(\{n\}) = e^{-a}\,\frac{a^n}{n!} \qquad n = 0,1,2,\ldots$

then

(0.1.16) $p_n(x) = (a^n n!)^{-1/2}\, C^{(a)}_n(x)$

where $\{C^{(a)}_n(x) : n = 0,1,2,\ldots\}$ are the *Charlier polynomials*. Sometimes these polynomials are referred to as *Poisson-Charlier polynomials* because the spectral measure corresponds to the Poisson distribution. An explicit expression for the Charlier polynomials is

(0.1.17) $C^{(a)}_n(x) = \sum_{j=0}^{n} \binom{n}{j}\binom{x}{j} j!\,(-a)^{n-j}$.

Example 5 (Chihara [39], p. 175).

For the discrete measure $(\beta > 0,\ 0 < c < 1)$

(0.1.18) $\mu(\{n\}) = (1-c)^{\beta}\,(\beta)_n\,\frac{c^n}{n!} \qquad n = 0,1,2,\ldots$

(which is the Pascal distribution or the negative-binomial distribution) the orthogonal polynomials are

$$(0.1.19)\qquad p_n(x) = \left(\frac{c^n}{n!}\right)^{1/2} m_n(x;\beta,c)$$

with $\{m_n(x;\beta,c) : n = 0,1,2,\dots\}$ the *Meixner polynomials*, given by

$$(0.1.20)\qquad m_n(x;\beta,c) = n! \sum_{j=0}^{n} \binom{x}{j}\binom{n+\beta-1}{j+\beta-1}\left(1-\frac{1}{c}\right)^j .$$

Chihara [39] refers to these polynomials as the *Meixner polynomials of the first kind*. The *Meixner polynomials of the second kind* (in Chihara's terminology) have the weight function

$$(0.1.21)\qquad w(x) = C\left|\Gamma\left(\frac{\eta+ix}{2}\right)\right|^2 \exp(-x \operatorname{Arctan} \delta)$$

with C a normalizing constant. Askey and Wilson [9] refer to these polynomials as the *Meixner–Pollaczek* polynomials.

Example 6 (Chihara [39], p. 184)

Consider the weight function ($a \geq |b|$, $\lambda > 0$)

$$(0.1.22)\qquad w(x) = \begin{cases} \dfrac{2^{2\lambda}(2\lambda+a)}{4\pi\,\Gamma(2\lambda)}(1-x^2)^{\lambda-1/2} \exp\left(-\operatorname{Arcsin} x \dfrac{ax+b}{\sqrt{1-x^2}}\right) & \\ \qquad \times \left|\Gamma\left(\lambda + i\dfrac{ax+b}{2\sqrt{1-x^2}}\right)\right|^2 & -1 < x < 1 \\ 0 & \text{elsewhere} \end{cases}$$

then the orthogonal polynomials are given by

$$(0.1.23)\qquad p_n(x) = \left\{\frac{n!\,(2n+2\lambda+a)}{(2\lambda)_n\,(2\lambda+a)}\right\}^{1/2} P_n^\lambda(x;a,b)$$

where $\{P_n^\lambda(x;a,b) : n = 0,1,2,\dots\}$ are the *Pollaczek polynomials* on $[-1,1]$. These can be written explicitely as

$$(0.1.24)\qquad n!\,P_n^\lambda(x;a,b) = (2\lambda)_n\, e^{in\theta} \sum_{j=0}^{n} \binom{n}{j} \frac{(\lambda+it)_j}{(2\lambda)_j} (e^{-2i\theta}-1)^j$$

with $x = \cos\theta$ and $t = \dfrac{ax+b}{2\sqrt{1-x^2}}$.

There are many more sequences of orthogonal polynomials. More examples are given in Chihara's book [39]. In Chapter 1 (§ 1.4) there will be a short discussion about orthogonal polynomials with a singular spectral measure supported on a Cantor set.

0.2. General properties

We will now mention some general properties of orthogonal polynomials. We will not give proofs here but refer to the literature (Szegö [175], Freud [61], Chihara [39]).

Lemma 0.1. (Szegö [175], p. 44). The zeros of orthogonal polynomials are real and simple and belong to the interval (a,b), where a and b are respectively the infimum and supremum of supp(μ) (we take $a = -\infty$ and/or $b = \infty$ when these do not exist). If we order the zeros of p_n in such a way that

$$a < x_{1,n} < x_{2,n} < \dots < x_{n,n} < b \tag{0.2.1}$$

then $x_{j,n+1} < x_{j,n} < x_{j+1,n+1}$ $(j = 1,\dots,n)$, which means that the zeros of p_n and p_{n+1} interlace.

The zeros of orthogonal polynomials serve very well as nodes for a numerical quadrature formule. They give rise to the *Gauss-Jacobi quadrature* :

Lemma 0.2. (Szegö [175], p. 47). Let

$$\lambda_{j,n} = \frac{k_{n+1}}{k_n} \frac{-1}{p_{n+1}(x_{j,n})p_n'(x_{j,n})} = \frac{k_n}{k_{n-1}} \frac{1}{p_{n-1}(x_{j,n})p_n'(x_{j,n})} \tag{0.2.2}$$

then for every polynomial π of degree less than or equal to 2n-1

$$\int_{-\infty}^{\infty} \pi(x)d\alpha(x) = \sum_{j=0}^{n} \lambda_{j,n}\pi(x_{j,n}) . \tag{0.2.3}$$

The numbers $\{\lambda_{j,n} : j = 1,\dots,n\}$ are called *Christoffel numbers* and they are all positive.

One of the most important properties of orthogonal polynomials is the existence of a recurrence relation for three consecutive polynomials :

Lemma 0.3. (Szegö [175], p. 42; Freud [61], p. 60). Orthogonal polynomials always satisfy a three term recurrence relation

$$(0.2.4)\qquad xp_n(x) = a_{n+1}p_{n+1}(x) + b_np_n(x) + a_np_{n-1}(x) \qquad n = 0,1,2,\ldots$$

with starting values $p_{-1}(x) = 0$ and $p_0(x) = 1$. The recurrence coefficients are given by

$$(0.2.5)\qquad \begin{cases} a_n = \dfrac{k_{n-1}}{k_n} > 0 & n = 1,2,\ldots \\ b_n = \displaystyle\int_{-\infty}^{\infty} xp_n^2(x)d\mu(x) \in \mathbb{R} & n = 0,1,2,\ldots\ . \end{cases}$$

If, on the other hand, $\{p_n(x) : n = 0,1,2,\ldots\}$ is a sequence of polynomials that satisfies a recurrence relation of the form (0.2.4) with $a_n > 0$ and $b_{n-1} \in \mathbb{R}$ $(n = 1,2,\ldots)$, then there exists a probability measure μ such that these polynomials are orthogonal with spectral measure μ.

The second part of the previous lemma is often referred to as *Favard's theorem*. The recurrence relation for the monic polynomials becomes

$$(0.2.6)\qquad \hat{p}_{n+1}(x) = (x - b_n)\hat{p}_n(x) - a_n^2\hat{p}_{n-1}(x) \qquad n = 0,1,2,\ldots\ .$$

The recurrence relation always implies the existence of a spectral measure with respect to which the polynomials are orthogonal. This measure need not be unique. The measure will be unique if and only if the *Hamburger moment problem* associated with this measure has a unique solution. A sufficient condition has been given by Carleman, namely

$$(0.2.7)\qquad \sum_{n=1}^{\infty} \frac{1}{a_n} = \infty$$

(Shohat-Tamarkin [170], p. 59).

The recurrence relation and the Gauss-Jacobi quadrature have some consequences. Define the *associated polynomials* $\{p_n^{(1)}(x) : n = 0,1,2,\ldots\}$ by means of the recurrence formula

$$(0.2.8)\qquad xp_n^{(1)}(x) = a_{n+2}p_{n+1}^{(1)}(x) + b_{n+1}p_n^{(1)}(x) + a_{n+1}p_{n-1}^{(1)}(x)$$

with $p_{-1}^{(1)}(x) = 0$ and $p_0^{(1)}(x) = 1$, then the decomposition into partial fractions of the ratio $p_{n-1}^{(1)}(x)/p_n(x)$ is given by

(0.2.9) $$\frac{p_{n-1}^{(1)}(x)}{p_n(x)} = a_1 \frac{\hat{p}_{n-1}^{(1)}(x)}{\hat{p}_n(x)} = a_1 \sum_{j=1}^{n} \frac{\lambda_{j,n}}{x - x_{j,n}}$$

which means that the Christoffel numbers are the residues of the ratio $\hat{p}_{n-1}^{(1)}(x)/\hat{p}_n(x)$. Another important rational function is

(0.2.10) $$\frac{\hat{p}_{n-1}(x)}{\hat{p}_n(x)} = \sum_{j=1}^{n} \frac{a_{j,n}}{x - x_{j,n}}$$

with

(0.2.11) $$a_{j,n} = \frac{\hat{p}_{n-1}(x_{j,n})}{\hat{p}_n'(x_{j,n})} = \lambda_{j,n} p_{n-1}^2(x_{j,n}) > 0 \ .$$

If the recurrence relation (0.2.4) is known, then one can introduce the *Jacobi matrix* (of order n) for the orthogonal polynomials :

(0.2.12) $$J_n = \begin{bmatrix} b_0 & a_1 & 0 & 0 & \cdots & & \\ a_1 & b_1 & a_2 & 0 & \cdots & & \\ 0 & a_2 & b_2 & a_3 & \cdots & & \\ \vdots & \vdots & & \ddots & \ddots & \ddots & \\ \vdots & \vdots & & & a_{n-2} & b_{n-2} & a_{n-1} \\ \vdots & \vdots & & & 0 & a_{n-1} & b_{n-1} \end{bmatrix}$$

The eigenvalues of this matrix are equal to the zeros of p_n; this follows immediately by expanding the determinant of $J_n - xI$ along the last row. A normalized eigenvector for the eigenvalue $x_{j,n}$ is given by

$$\sqrt{\lambda_{j,n}}\ (p_0(x_{j,n}), p_1(x_{j,n}), \ldots, p_{n-1}(x_{j,n}))$$

where $\{\lambda_{j,n} : j = 1,\ldots,n\}$ are the Christoffel numbers (0.2.2). The monic orthogonal polynomials therefore can be written as

(0.2.13) $$\hat{p}_n(x) = \det(xI - J_n) \ .$$

Another matrix representation for the orthogonal polynomials is given in terms of the moments (0.1.1) :

$$(0.2.14) \qquad p_n(x) = \frac{1}{\sqrt{D_n D_{n-1}}} \begin{vmatrix} m_0 & m_1 & m_2 & \cdots & m_n \\ m_1 & m_2 & m_3 & \cdots & m_{n+1} \\ \vdots & & & & \\ m_{n-1} & m_n & m_{n+1} & \cdots & m_{2n-1} \\ 1 & x & x^2 & \cdots & x^n \end{vmatrix}$$

where D_n is the determinant of the *Hankel matrix* H_n with elements $\{(H_n)_{i,j} = m_{i+j} : i,j = 0,1,2,\ldots,n\}$.

We can obtain some more interesting formulas from the recurrence formula :

Lemma 0.4. (Szegö [175], p. 43). The following formulas always hold for orthogonal polynomials :

$$(0.2.15) \qquad \sum_{j=0}^{n} p_j(x)p_j(y) = \frac{k_n}{k_{n+1}} \frac{p_{n+1}(x)p_n(y) - p_n(x)p_{n+1}(y)}{x - y}$$

$$(0.2.16) \qquad \sum_{j=0}^{n} p_j^2(x) = \frac{k_n}{k_{n+1}} \{p'_{n+1}(x)p_n(x) - p'_n(x)p_{n+1}(x)\} .$$

The first equality is called the *Christoffel-Darboux formula*, the second formula is a confluent form of the Christoffel-Darboux formula.

We finally mention the following minimal property :

Lemma 0.5. (Szegö [175], p. 39). Let Π_n^+ be the set of all polynomials with leading term x^n, then

$$(0.2.17) \qquad \frac{1}{k_n^2} = \inf_{q_n \in \Pi_n^+} \int_{-\infty}^{\infty} |q_n(x)|^2 \, d\mu(x)$$

and the infimum is attained by the monic orthogonal polynomial $\hat{p}_n$ with spectral measure μ.

0.3. Outline of this monograph

In this monograph we study orthogonal polynomials for which the degree increases to infinity. In Chapter 1 we discuss the asymptotic behaviour of orthogonal polynomials on a compact set. We give the connection with the Green's function of the complement of this compact set and the equilibrium measure (from logarithmic potential theory) for this set. We show that the equilibrium measure gives the asymptotic distribution of the zeros of the orthogonal polynomials. More precise results are given for orthogonal polynomials on the interval [-1,1], especially those that belong to the Szegö class, and a class of orthogonal polynomials on Cantor sets is discussed briefly.

In Chapter 2 we deal with the recurrence relation for orthogonal polynomials. We give results for orthogonal polynomials when the recurrence coefficients are asymptotically periodic. We construct the spectral measure for these polynomials on and off the spectrum. This spectrum consists of a disjoint union of several intervals and at most a denumerable set of mass points. Special attention is paid to the case where the period is one, in which case the polynomials are orthogonal on one interval.

In Chapter 3 a new method is given, to obtain asymptotic formulas for sequences of polynomials. This method is based on well-known theorems from probability theory. Some classical (and semi-classical) orthogonal polynomials will be handled and asymptotic expansions are given, valid for large values of the degree.

Orthogonal polynomials on infinite intervals are studied in Chapter 4. We discuss recent results involving the zero distribution for orthogonal polynomials with exponential weights. Asymptotic results for the largest zeros, for the leading coefficient and Plancherel-Rotach asymptotics for some orthogonal polynomials are given. We also introduce weighted zero distributions and develop the asymptotic theory for these.

In Chapter 5 we give some consequences of the existence of the asymptotic zero distribution. We show how the asymptotic behaviour of the coefficients of orthogonal polynomials, weak asymptotics for Christoffel functions and the distribution of the eigenvalues of Toeplitz matrices associated with orthogonal polynomials can be obtained from the asymptotic zero behaviour. Functions of the second kind are discussed briefly.

In Chapter 6 we give some applications of the theory given in the other chapters. We mention the connection with the distribution of the eigenvalues of random matrices, we discuss the physical interpretation of the recurrence relation as a discrete Schrödinger equation and the probabilistic interpretation in the framework of birth-and-death processes.

0.4. A short historical note

A general theory of orthogonal polynomials has started with the investigation of a special type of continued fraction. The first works on the subject are those of the Russian mathematician *Pafnuti Lvovich Chebyshev* (1821-1894) [38] and the Dutch mathematician *Thomas Jan Stieltjes* (1856-1894) [174] who, independent of each other, proved a number of properties valid for general orthogonal polynomials. The concept of orthogonality with respect to a general distribution (measure) seems to be due to Stieltjes [173].

Special systems of orthogonal polynomials were known before that time. We can go back in history until the classical memoir of *Adrien Marie Legendre* (1752-1833) on the motion of planets [107], published in 1785, where the polynomials, that now bear his name, were mentioned. *Joseph-Louis Lagrange* (1736-1813) already came across the recurrence relation for these Legrendre polynomials earlier [104']. General Jacobi polynomials appeared in 1859 in a paper by *Carl Jacob Jacobi* (1804-1851) [90]. Laguerre polynomials ($\alpha = 0$) already appear in works of *Niels Henrik Abel* (1802-1829), *Joseph-Louis Lagrange* (1736-1813) and Chebyshev [37] even before *Edmond Nicolas Laguerre* (1834-1886) treated them in 1879 [105]. Generalized Laguerre polynomials were first studied by *Yulian-Karl Sokhotsky* (1842-1927) and later by *Nikolai Yakovlevich Sonin* (1849-1915) [171]. Hermite polynomials appear for the first time in a mémoire by *Pierre Simon de Laplace* (1749-1827) [106] and were also considered by Chebyshev [37] before *Charles Hermite* (1822-1901) used them [84]. The Charlier polynomials were introduced by *Carl Wilhelm Ludwig Charlier* (1862-1934) [36]. Meixner polynomials were studied by *Josef Meixner* [131] who was investigating a special type of generating functions. Meixner polynomials were also considered by *Ervin Feldheim* [59], independent of Meixner, and even Stieltjes mentions them briefly [174]. *Felix Pollaczek* (1892-1981) described the polynomials with his name in a series of papers [158] - [160]. The latter polynomials are an important example of orthogonal polynomials for which Szegö's theory for orthogonal polynomials on [-1,1] is not valid.

The quadrature formula (0.2.3) was first given by *Karl Friedrich Gauss* (1777-1855) for Legendre polynomials [68] and Jacobi's proof for this case [89] is still being used in textbooks. The proof for general distributions was given by Stieltjes [173]. The Christoffel-Darboux formula was already published by Chebyshev [38'] before *Elwin Bruno Christoffel* (1829-1900) discovered it for Legendre polynomials [46] and later for the general case about the same time as *Jean Gaston Darboux* (1842-1917) [47]. Favard's theorem was announced by *Jean Aimé Favard* [58] but was already explicitely given by Stone [174'] and in a disguised form by Perron [153].

The classical orthogonal polynomials were well studied and a lot of properties were found, in particular the asymptotic behaviour was well known. A general theory for the asymptotics of orthogonal polynomials on the interval [-1,1] started with the investigations of *Sergei Natanovich Bernstein* (1880-1968) [22] and culminated in the theory developped by *Gabor Szegö* (1895-1985). Szegö introduced orthogonal polynomials on the unit circle [176], [177] and emphasized the close connection with ortogonal polynomials on [-1,1]. Szegö only treated absolutely continuous measures; his theory was generalized to more general measures by *Andrei Nikolaevich Kolmogorov* (1903-) [102] and *Mark Grigorievich Krein* (1907-) [104]. One also should mention *Géza Freud* (1922-1979) [61] - [62] and *Yakov Lazonovich Geronimus* [75] - [77] who made important contributions to Szegö's theory.

The study of orthogonal polynomials lay dormant at the beginning of this century. The publication of Szegö's book [175] helped to revive the interest in orthogonal polynomials. This book is still the best book about orthogonal polynomials, even though its first edition goes back as far as 1939. During the last decades the asymptotic theory for orthogonal polynomials on noncompact sets was initiated by Freud [63] - [67] and his investigations had a lot of influence on contemporary mathematicians as *Paul Nevai* (and his collaborators) [138] - [151] and *Joseph Ullman* [180] - [185]. *Ken Case* and *Marc Kac* showed that some aspects of the theory of orthogonal polynomials were closely connected to (discrete) scattering theory [35], an observation that led to new and interesting methods of proof [34], [69] - [74]. *Richard Askey* has done a lot of research on orthogonal polynomials that can be written as special functions [8]. He and many other mathematicians have become very interested in orthogonal polynomials that can be written as q-hypergeometric (or basic hypergeometric) functions. These polynomials will not be handled in this monograph but we refer to an excellent account of this research by Askey and Wilson [9].

CHAPTER 1 : ORTHOGONAL POLYNOMIALS ON A COMPACT SET

1.1. Some notions from potential theory

In this section we will introduce some notions from (logarithmic) potential theory. Let E be a compact set in $\mathbb{R}$ and Ω_E the class of all probability measures on E. If μ is a measure that belongs to Ω_E, then its *energy* is defined as

$$(1.1.1) \qquad I(\mu) = \int_E \int_E \log \frac{1}{|x-y|} \, d\mu(x) d\mu(y) .$$

Since E is compact there exists a constant K such that $|x-y| \leq K$ $(x,y \in E)$ whence $I(\mu) \geq -\log K$. The (logarithmic) *potential* of the measure μ is

$$(1.1.2) \qquad U(z;\mu) = \int_E \log \frac{1}{|z-x|} \, d\mu(x) \qquad z \in \mathbb{C}$$

and with the same reasoning as above it follows that $U(z;\mu) \geq -\log K$ for $z \in E$. The *equilibrium energy* on E (*Robin's constant*) is then given by

$$(1.1.3) \qquad V(E) = \inf_{\mu \in \Omega_E} I(\mu)$$

and the (logarithmic) *capacity* of E is

$$(1.1.4) \qquad C(E) = e^{-V(E)} .$$

The capacity for an arbitrary bounded set B in $\mathbb{R}$ is defined as

$$C(B) = \sup\{C(K) : K \subset B, \ K \text{ compact}\} .$$

A well-known result from potential theory is that for every compact set E with positive capacity $C(E) > 0$, there exists a unique measure $\mu_E \in \Omega_E$ such that

$$(1.1.5) \qquad V(E) = I(\mu_E) .$$

This measure is the *equilibrium measure* (or *Frostman measure*) for E. It can be characterized as being the only measure in Ω_E for which

$$U(x;\mu_E) \le \log \frac{1}{C(E)} = V(E) \quad , \quad x \in \mathbb{C} \tag{1.1.6}$$

and equality holds if $x \in E\backslash B$, with B a Borel set of type F_σ (a countable union of compact sets) and capacity zero (Hille [85], p. 282). All the definitions so far still make sense when E is a compact set in $\mathbb{C}$, but we will always take E real since we only consider orthogonal polynomials on the real line.

Let $\{\mu_n : n = 1,2,3,...\}$ be a sequence of probability measures on $\mathbb{R}$, then μ_n *converges weakly* to a probability measure μ if for every bounded and continuous function f on $\mathbb{R}$

$$\lim_{n\to\infty} \int_{-\infty}^{\infty} f(x)d\mu_n(x) = \int_{-\infty}^{\infty} f(x)d\mu(x) \ ,$$

and this is denoted by $\mu_n \to \mu$. The next lemma gives a sufficient condition for a sequence of probability measures to converge weakly to the equilibrium measure of a compact set E :

Lemma 1.1. (Brolin [32], p. 140). Let E and H be two compact sets, $E \subset H$ and $C(E) > 0$. Suppose $\{\mu_n : n = 1,2,...\}$ is a sequence of probability measures on H that converges weakly to a probability measure $\mu \in \Omega_E$. If

(a) $\liminf_{n\to\infty} U(x;\mu_n) \ge \log \frac{1}{C(E)} \qquad (x \in E\backslash B,\ \mu_E(B) = 0)$

(b) $\operatorname{supp}(\mu_E) = E$

then the weak limit μ is the equilibrium measure μ_E.

Proof : From (a) it follows that

$$\log \frac{1}{C(E)} \le \int_E \liminf_{n\to\infty} U(x;\mu_n)d\mu_E(x) \ . \tag{1.1.7}$$

On the other hand Fubini's theorem implies

$$\int_E U(x;\mu_n)d\mu_E(x) = \int_H U(x;\mu_E)d\mu_n(x)$$

and Fatou's lemma, together with (1.1.6) yields

$$\int_E \liminf_{n\to\infty} U(x;\mu_n)d\mu_E(x) \le \liminf_{n\to\infty} \int_H U(x;\mu_E)d\mu_n(x) \tag{1.1.8}$$

$$\le \log \frac{1}{C(E)} \ .$$

This means that

$$\liminf_{n \to \infty} U(x;\mu_n) = \log \frac{1}{C(E)} \qquad x \in E \backslash B_1$$

with $\mu_E(B_1) = 0$. The lower envelope theorem (de la Vallée Poussin [186], p. 85) implies that

$$U(x;\mu) \leqslant \liminf_{n \to \infty} U(x;\mu_n) \qquad x \in E$$

with equality on $E \backslash B_2$ and $C(B_2) = 0$, hence

$$U(x;\mu) \leqslant \log \frac{1}{C(E)} \qquad x \in E \backslash B_1 \quad , \quad \mu_E(B_1) = 0 \ .$$

From (b) it follows that every neighbourhood of a point $z \in E$ contains points for which $U(x;\mu) \leqslant \log \frac{1}{C(E)}$. The potential of a measure is a lower semi-continuous function, hence

$$U(z;\mu) \leqslant \liminf_{x \to z} U(x;\mu)$$

from which we find that $U(x;\mu) \leqslant \log \frac{1}{C(E)}$ for every $x \in E$. Since μ_E is the only measure with this property we conclude that $\mu = \mu_E$. ∎

Denote by $\bar{\mathbb{C}}$ the compactified complex plane $\mathbb{C} \cup \{\infty\}$. Let E be a compact set in $\mathbb{R}$ with positive capacity and $\text{supp}(\mu_E) = E$ and let $G = \bar{\mathbb{C}} \backslash E$. If E consists of a finite number of disjoint intervals then the *Green's function* of G (with pole at ∞) is the unique function g_E that is harmonic in $G \backslash \{\infty\}$ with a behaviour at ∞ given by

$$(1.1.9) \qquad g_E(z) = \log|z| + \text{harmonic function}$$

and for which

$$(1.1.10) \qquad \lim_{\substack{z \to x \\ z \in G}} g_E(z) = 0 \qquad x \in E \ .$$

In the general case we let $G = \bigcup_{n=1}^{\infty} G_n$ with $\{G_n : n = 1,2,\dots\}$ an increasing family of sets for which the complement is equal to a finite number of disjoint intervals. The corresponding sequence of Green's functions is increasing and the limit defines the Green's function for G (Tsuji [179], p. 14-19). The Green's function is always positive and harmonic in $G \backslash \{\infty\}$ and behaves near ∞ as in (1.1.9), but (1.1.10) will not necessarily be true for every $x \in E$. The relation between the Green's function and the capacity is

(1.1.11) $$\lim_{z \to \infty} \{g_E(z) - \log|z|\} = \log \frac{1}{C(E)}$$

and the Green's function will exist if and only if $C(E) > 0$ (Tsuji [179], p. 81). One can easily verify that

(1.1.12) $$g_E(z) = -U(z;\mu_E) - \log C(E)$$

satisfies all the requirements for the Green's function of G, and (1.1.10) will be true for every $x \in E\backslash B$, with $C(B) = 0$ because of (1.1.6). Let $\hat{g}_E(z) = g_E(z) + ih(z)$, where h is a conjugate harmonic function chosen in such a way that $\hat{g}_E$ is analytic in G except at ∞ where $\hat{g}_E(z) - \log z$ is analytic, then $\operatorname{Re} \hat{g}_E(z) = 0$ for $z \in E$, with the exception of a set of capacity zero. The function

$$\omega_E(z) = \exp \hat{g}_E(z)$$

maps G into the component of the complement of the unit circle $\{z \in \mathbb{C} : |z| = 1\}$ that contains ∞, but this mapping is in general not a one-to-one mapping since G need not be simply connected. If z is large enough then $\omega(z)$ will be a conformal mapping and the inverse mapping γ can be defined. Now define for every $\theta \in [0,2\pi)$ the number $r_0(\theta)$ as being the infimum of all numbers $r_0 \geq 1$ for which γ can be analytically continued from ∞ along $\{re^{i\theta} : r > r_0\}$. Let

$$R_\theta = \{re^{i\theta} : r > r_0(\theta)\}$$

then

$$S = \bigcup_{0 \leq \theta < 2\pi} \gamma(R_\theta)$$

is the *Green's star domain* for G (Sario and Nakai [168]). The function ω is then conformal on S and maps S to the exterior of the unit circle minus radial segments emanating from the unit circle.

Finally we define the *Chebyshev polynomials* on E as the monic polynomials $\{T_n(x;E) : n = 0,1,2,...\}$ for which

(1.1.13) $$M_n = \sup_{x \in E} |T_n(x;E)| = \inf_{q_n \in \Pi_n^+} \sup_{x \in E} |q_n(x)|$$

with Π_n^+ as in Lemma 0.5. These polynomials are unique and have the property that

(1.1.14) $$\lim_{n \to \infty} M_n^{1/n} = C(E)$$

(Tsuji [179], p. 72-73). Notice that for $E = [-1,1]$ the Chebyshev polynomials $\{T_n(x;E) : n = 0,1,2,\ldots\}$ correspond to the Chebyshev polynomials of the first kind $\{2^{n-1}T_n(x) : n = 0,1,2,\ldots\}$.

1.2. Main theorem

In this section we investigate the zero behaviour of orthogonal polynomials on a compact set. Introduce the measures $\{\nu_n : n = 1,2,\ldots\}$ by

$$(1.2.1) \qquad \begin{cases} \nu_n(\{x_{j,n}\}) = \frac{1}{n} & j = 1,2,\ldots,n \\ \nu_n(A) = 0 & A \text{ contains no zeros of } p_n(x;\mu) \end{cases}$$

where $\{p_n(x;\mu) = k_n x^n + \ldots\}$ are the orthogonal polynomials with spectral measure μ and $\{x_{j,n}\}$ are the zeros of p_n.

Theorem 1.2. (Van Assche [190]) Let E be a compact set in $\mathbb{R}$ with positive capacity and $\text{supp}(\mu_E) = E$, and E^* a bounded denumerable set with accumulation points in E. Suppose μ is a probability measure on $E \cup E^*$ such that $\mu = \mu_1 + \mu_2$, with μ_1 absolutely continuous and μ_2 singular with respect to the equilibrium measure μ_E. Then $\mu_E(\{d\mu_1/d\mu_E > 0\}) = 1$ implies

$$(1.2.2) \qquad \lim_{n \to \infty} k_n^{-1/n} = C(E)$$

and

$$(1.2.3) \qquad \nu_n \Rightarrow \mu_E \qquad (n \to \infty) .$$

The theorem as it stands here is a somewhat weaker version of a theorem by Widom [201] who imposes some "admissibility" condition for the spectral measure μ and the equilibrium measure μ_E. A first result of this type was proved by Erdös and Turán [54] for the case $E = [-1,1]$ and $\mu_2 = 0$. Similar investigations can be found in Ullman [180] - [184]. The weak convergence in (1.2.3) means that

$$\lim_{n \to \infty} \frac{1}{n} \sum_{j=1}^{n} f(x_{j,n}) = \int_E f(x)d\mu_E(x)$$

for every bounded and continuous function f on $\mathbb{R}$. This is a result on the asymptotic distribution of the zeros of orthogonal polynomials that tells us that the zeros eventually are distributed according to the measure μ_E. The basic idea of the proof was inspired by Ullman [180] and consists in showing that the L_2-norm of the sequence of orthogonal polynomials behaves the same as the L_∞-norm of the sequence of Chebyshev polynomials on $E \cup E^*$. This will furnish some information concerning the zero behaviour of the orthogonal polynomials. We need some lemma's :

Lemma 1.3. (Ullman [182]) : Let μ be a positive measure on a set $E \subset \mathbb{R}$ and $\{f_n(x) : n = 1,2,\ldots\}$ a sequence of non-negative μ-measurable functions on E. Suppose that $\{m_n : n = 1,2,\ldots\}$ is an increasing sequence of positive integers such that

$$\limsup_{n\to\infty} \{\int_E f_{m_n}(x)d\mu(x)\}^{1/m_n} \leq 1$$

then there exists a subsequence $\{t_n : n = 1,2,\ldots\}$ of $\{m_n : n = 1,2,\ldots\}$ and a Borel set B with $\mu(B) = 0$ such that

$$\limsup_{n\to\infty} (f_{t_n}(x))^{1/t_n} \leq 1 \qquad x \in E\backslash B$$

Proof : Let $\{\varepsilon_n : n = 1,2,\ldots\}$ be a positive sequence for which $\varepsilon_n \longrightarrow 0$ and

$$\{\int_E f_{m_n}(x)d\mu(x)\}^{1/m_n} \leq 1 + \varepsilon_{m_n} .$$

Let $g_n(x) = f_n(x)/\{n(1+\varepsilon_n)^n\}$ then it follows immediately that g_{m_n} converges to zero in the $L_1(\mu)$-norm. This implies that there exists a subsequence $\{t_n : n = 1,2,\ldots\}$ of $\{m_n : n = 1,2,\ldots\}$ such that g_{t_n} converges to zero almost everywhere with respect to the measure μ, which means that there exists a Borel set $B \subset E$ with $\mu(B) = 0$ such that

$$\lim_{n\to\infty} g_{t_n}(x) = 0 \qquad x \in E\backslash B .$$

Consequently, for every $x \in E\backslash B$ there exists an integer N (which may depend on x) such that $g_{t_n}(x) \leq 1$ for $n > N$, which is the same as

$$f_{t_n}(x) \leq t_n(1 + \varepsilon_{t_n})^{t_n} \qquad n > N \quad , \quad x \in E\backslash B ,$$

from which the result follows. ∎

Lemma 1.4. : Let $\{\nu_n : n = 1,2,\ldots\}$ be the sequence of measures defined in (1.2.1). If the spectral measure μ is such that $\text{supp}(\mu) = E \cup E^*$, where E is a compact set and all the accumulation points of E^* are in E, then every subsequence of

$\{\nu_n : n = 1,2,\ldots\}$ that converges weakly has a weak limit with support in E.

Proof : Choose $\varepsilon > 0$ and define

$$E^\varepsilon = \bigcup_{x \in E} [x-\varepsilon, x+\varepsilon] .$$

The union actually contains only a finite number of sets because E is compact and therefore E^ε is also compact. The set $\mathbb{R} \setminus E^\varepsilon$ is open and can be written as a denumerable union of open intervals $\{A_j : j = 1,2,\ldots\}$. Because of the conditions it follows that $A_j \cap \mathrm{supp}(\mu)$ contains at most a finite number of points for every j. Therefore $\{p_n(x) : n = 0,1,2,\ldots\}$ can have at most a finite number of zeros in A_j (Szegö [175], p. 50) so that

$$\nu_n(A_j) = O(\tfrac{1}{n}) \qquad j = 1,2,3,\ldots .$$

Use the portmanteau theorem (Billingsley [28], p. 11-12) to conclude that

$$\nu(A_j) \leq \liminf_{n\to\infty} \nu_{m_n}(A_j) = 0 \qquad j = 1,2,\ldots$$

where ν is the weak limit of the converging subsequence $\{\nu_{m_n} : n = 1,2,\ldots\}$. Hence σ-additivity implies that $\nu(\mathbb{R} \setminus E^\varepsilon) = 0$ so that the support of ν is a part of E^ε. This being true for every $\varepsilon > 0$ gives the desired result. ∎

Now we can prove the main theorem. We will do this in two steps : first we prove the asymptotic behaviour of the leading coefficient given by (1.2.2) and then we use that result to establish the weak convergence in (1.2.3).

Proof of Theorem 1.2. : From the minimal property (0.2.17) we get

$$\frac{1}{k_n^2} \leq \int_{E \cup E^*} |T_n(x;E \cup E^*)|^2 d\mu(x)$$

where $T_n(x;E \cup E^*)$ is the Chebyshev polynomial of degree n for the set $E \cup E^*$. Consequently

$$\frac{1}{k_n^2} \leq M_n^2 = \sup_{x \in E \cup E^*} |T_n(x;E \cup E^*)|^2$$

from which

$$\text{(1.2.4)} \qquad \limsup_{n\to\infty} k_n^{-1/n} \leq \lim_{n\to\infty} M_n^{1/n} = C(E \cup E^*) = C(E) ,$$

the last equality holds since E^{*} is denumerable so that $C(E^{*}) = 0$ (Tsuji [179], p. 57, 63).

Suppose now that $k_n^{-1/n}$ does not converge to $C(E)$, which means that

$$\liminf_{n \to \infty} k_n^{-1/n} \leq \alpha < C(E) .$$

In that case there exists an increasing sequence $\{m_n : n = 1,2,\ldots\}$ such that

$$\lim_{n \to \infty} \left(\int |\hat{p}_{m_n}(x;\mu)|^2 d\mu(x) \right)^{1/m_n} \leq \alpha^2$$

and if we set $\mu = \mu_1 + \mu_2$ then it follows that

$$\limsup_{n \to \infty} \left(\int_E |\hat{p}_{m_n}(x;\mu)|^2 w(x) d\mu_E(x) \right)^{1/m_n} \leq \alpha^2$$

where w is the Radon-Nikodym derivative $d\mu_1/d\mu_E$. Use Lemma 1.3 with $f_n(x) = \alpha^{-2n} |\hat{p}_n(x;\mu)|^2 w(x)$ and the fact that $w(x) > 0$ almost everywhere with respect to the measure μ_E to find a subsequence $\{t_n : n = 1,2,\ldots\}$ such that

$$(1.2.5) \qquad |\hat{p}_{t_n}(x;\mu)|^{1/t_n} \leq \alpha \qquad\qquad x \in E \backslash B_1$$

with $\mu_E(B_1) = 0$. The sequence $\{\nu_n : n = 1,2,\ldots\}$ consists of measures with support on the convex hull of $E \cup E^{*}$, which is a compact set, so that there always exist subsequences that converge weakly to a probability measure. Let $\{s_n : n = 1,2,\ldots\}$ be a subsequence of $\{t_n : n = 1,2,\ldots\}$ such that $\nu_{s_n} \to \nu$ $(n \to \infty)$, then the support of the weak limit ν will be a part of E (Lemma 1.4) and from the lower envelope theorem (de la Vallée Poussin [186], p. 85) we deduce

$$(1.2.6) \qquad U(x;\nu) \leq \liminf_{n \to \infty} U(x;\nu_{s_n}) \qquad\qquad x \in E$$

where the inequality holds only on a subset $B_2 \subset E$ with capacity zero. For this subset one also finds that $\mu_E(B_2) = 0$ (Tsuji [179], p. 56). Clearly

$$U(x;\nu_n) = \log |\hat{p}_n(x;\mu)|^{-1/n}$$

so that combining (1.2.5) and (1.2.6) leads to

$$U(x;\nu) \geq \log \frac{1}{\alpha} > \log \frac{1}{C(E)} \qquad\qquad x \in E \backslash (B_1 \cup B_2) .$$

Since $\mu_E(B_1 \cup B_2) = 0$ and $\text{supp}(\mu_E) = E$, we can use Lemma 1.1 from which we conclude that $\nu = \mu_E$. But then $U(x;\nu)$ will be equal to $\log \frac{1}{C(E)}$ for $x \in E \backslash B_3$ with $C(B_3) = 0$. This gives a contradiction, from which we conclude that $k_n^{-1/n}$ converges to $C(E)$.

Next we will show that ν_n converges weakly to μ_E. Repeat the previous reasoning but set $\alpha = C(E)$ everywhere. The conclusion is that for every subsequence of $\{\nu_n : n = 1,2,\ldots\}$ there exists a further subsequence that converges weakly to μ_E, from which the weak convergence of the sequence follows.

∎

Theorem 1.2 gives a sufficient condition, based on a comparison of the spectral measure μ with the equilibrium measure μ_E, under which the zeros of $\{p_n(x;\mu)\}$ are distributed according to μ_E. The condition is not a necessary one : Ullman [180] has constructed a sequence of orthogonal polynomials on [-1,1] for which the zeros behave regularly, i.e. $\nu_n \Rightarrow \mu_{[-1,1]}$, but for which the spectral measure μ is singular with respect to $\mu_{[-1,1]}$. Other counterexamples are given by Delyon-Simon-Souillard [49].

The asymptotic behaviour of the sequence $\{\nu_n : n = 1,2,3,\ldots\}$ has some immediate consequences for the asymptotics of the orthogonal polynomials. Define for the sequence $\{p_n(x;\mu) : n = 0,1,2,\ldots\}$ the sets

$$Z_N = \{x_{j,n} : 1 \leq j \leq n \ , \ n > N\} \ ,$$

$$X_1 = \{\text{accumulation points of } Z_0\} \ ,$$

$$X_2 = \{x \in Z_0 : p_n(x) = 0 \text{ for infinitely many } n\} \ .$$

These sets were introduced by Chihara [39]. Notice that for orthogonal polynomials on a compact set

$$\text{supp}(\mu) \subset X_1 \cup X_2 \subset \text{convex hull of supp}(\mu) \ .$$

<u>Corollary 1.5.</u> With the conditions of Theorem 1.2 we have

$$\lim_{n\to\infty} |p_n(z;\mu)|^{1/n} = \exp\{ \int_E \log |z-x| d\mu_E(x) - \log C(E)\} \tag{1.2.7}$$

$$= \exp g_E(z)$$

uniformly on compact subsets of $\mathbb{C}\backslash(X_1 \cup X_2)$. On E we have

$$\limsup_{n\to\infty} |p_n(z;\mu)|^{1/n} = 1$$

except possibly on a set B of type F_σ and capacity zero.

Proof : The polynomial $p_n(z;\mu)$ can be written in terms of the measure ν_n as

$$\frac{1}{n}\log|p_n(z;\mu)| = \frac{1}{n}\log k_n + \frac{1}{n}\sum_{j=1}^{n}\log|z - x_{j,n}|$$
$$= \frac{1}{n}\log k_n + \int \log|z - x|d\nu_n(x) .$$

If K is a compact set in $\mathbb{C}\backslash(X_1 \cup X_2)$ then there exists a positive integer N such that $K \cap (X_1 \cup X_2 \cup Z_N) = \phi$. For every $z \in K$ the function $f(x) = \log|z - x|$ will be continuous on $\mathrm{supp}(\nu_n)$ for $n \in N$ and (1.2.2) - (1.2.3) imply

$$\lim_{n\to\infty} \frac{1}{n}\log|p_n(z;\mu)| = \log\frac{1}{C(E)} + \int_E \log|z - x|d\mu_E(x) .$$

For every $z \in K$ we have

$$K_1 \leq |z - x_{j,n}| \leq K_2 \qquad n > N$$

with

$$K_1 = \inf\{|z - x| : z \in K, x \in X_1 \cup X_2 \cup Z_N\} > 0$$

$$K_2 = \sup_{z\in K} |z| + \sup_{x\in \mathrm{supp}(\mu)} |x| < \infty .$$

The quantity K_1 is strictly positive since K is compact, $X_1 \cup X_2 \cup Z_N$ is closed and $K \cap (X_1 \cup X_2 \cup Z_N) = \phi$. The constant K_2 is finite since K and $\mathrm{supp}(\mu)$ are both compact. We claim that the sequence $\{\frac{1}{n}\log|p_n(x;\mu)| : n > N\}$ is equicontinuous in K. Choose $\varepsilon > 0$, then there exists a $\delta > 0$ such that $|x - y| < \delta$ implies $|\log|x| - \log|y|| < \varepsilon$ provided that $x,y \in \{z \in \mathbb{C} : 0 < K_1 \leq |z| \leq K_2 < \infty\}$. Therefore if $|x - y| < \delta$ and $n > N$ we have

$$|\frac{1}{n}\log|p_n(x;\mu)| - \frac{1}{n}\log|p_n(y;\mu)||$$
$$\leq \frac{1}{n}\sum_{j=1}^{n}|\log|x - x_{j,n}| - \log|y - x_{j,n}|| < \varepsilon$$

from which the equicontinuity follows. The theorem of Arzèla-Ascoli then implies that $\{\frac{1}{n}\log|p_n(z;\mu)| : n > N\}$ converges uniformly to $g_E(z)$ on K.
On the set E we have by the lower envelope theorem and (1.1.6)

$$\limsup_{n\to\infty}\frac{1}{n}\log|\hat{p}_n(z;\mu)| = -\liminf_{n\to\infty} U(z;\nu_n)$$
$$= -U(z;\mu_E) = \log C(E)$$

except possibly on a set B of type F_σ and capacity zero, which proves the corollary. ∎

Corollary 1.6. With the conditions of Theorem 1.2 we have

$$\lim_{n\to\infty} \frac{1}{n}\frac{p_n'(z;\mu)}{p_n(z;\mu)} = \int_E \frac{1}{z-x}\, d\mu_E(x) \tag{1.2.8}$$

uniformly on compact subsets of $\mathbb{C}\backslash(X_1 \cup X_2)$.

Proof : If $z \in \mathbb{C}\backslash(X_1 \cup X_2)$ then for $n > N$ the function $f(x) = \frac{1}{z-x}$ will be continuous on $\mathrm{supp}(\nu_n)$. Since

$$\frac{1}{n}\frac{p_n'(z;\mu)}{p_n(z;\mu)} = \frac{1}{n}\sum_{j=1}^{n} \frac{1}{z - x_{j,n}} = \int \frac{1}{z-x}\, d\nu_n(x) \tag{1.2.9}$$

the asymptotic behaviour in (1.2.8) follows pointwise on $\mathbb{C}\backslash(X_1 \cup X_2)$. With the same notation as in the previous corollary we obtain the bound

$$\left|\frac{1}{n}\frac{p_n'(z;\mu)}{p_n(z;\mu)}\right| < \frac{1}{K_1} \qquad n > N$$

whenever z belongs to a compact set K in $\mathbb{C}\backslash(X_1 \cup X_2)$. Clearly $\frac{1}{n}\, p_n'(z;\mu)/p_n(z;\mu)$ is analytic in K and the uniform bound on K therefore implies that the convergence in (1.2.8) is uniform on K. ∎

1.3. Orthogonal polynomials on [-1,1]

If $E = [-1,1]$ then the capacity of E is 1/2 and the equilibrium measure on E is the so-called *arcsin measure* given by

$$\mu_E(A) = \frac{1}{\pi}\int_A \frac{dx}{\sqrt{1-x^2}} \tag{1.3.1}$$

with A a Borel set in [-1,1] (Ullman [180]). The Green's function is

$$g_E(z) = \log|z + \sqrt{z^2 - 1}| \tag{1.3.2}$$

where the square root is such that $|z + \sqrt{z^2-1}| > 1$ whenever $z \notin [-1,1]$. If the spectral measure is given by $\mu = w(x)dx + \mu_2$, where μ_2 is singular with respect to Lebesgue measure on [-1,1] with possible jumps outside [-1,1] that accumulate at ± 1 and $w(x) > 0$ almost everywhere in [-1,1], then it follows from Theorem 1.2 that $k_n^{-1/n}$ converges to 1/2 and ν_n converges weakly to the arcsin measure. In particular, this means that

$$\lim_{n\to\infty} \frac{1}{n} \sum_{j=1}^{n} f(x_{j,n}) = \frac{1}{\pi} \int_{-1}^{1} f(x) \frac{dx}{\sqrt{1-x^2}}$$

for every continuous function on [-1,1]. This result (without μ_2) was given by Erdős and Turán [54] and led to some interesting research by Ullman [180]. Corollary 1.5 thus yields

$$\lim_{n\to\infty} |p_n(x;\mu)|^{1/n} = |x + \sqrt{x^2-1}| \tag{1.3.3}$$

uniformly on compact subsets of $\mathbb{C}\backslash\mathrm{supp}(\mu)$, since for this case one obviously has $X_1 \cup X_2 = \mathrm{supp}(\mu) = [-1,1] \cup \{\text{mass points of } \mu_2 \text{ outside } [-1,1]\}$.

1.3.1. The Szegö class

Szegö developped a beautiful theory for orthogonal polynomials on [-1,1] that improves the asymptotic formula in (1.3.3) essentially. The theory works so well basically because $G = \bar{\mathbb{C}}\backslash[-1,1]$ is simply connected so that G can be mapped conformally to the exterior of the unit circle. Szegö introduced orthogonal polynomials on the unit circle, for which the theory of harmonic functions inside the unit circle could be used. There is an easy relation between orthogonal polynomials on the unit circle and orthogonal polynomials on [-1,1] which makes it possible to transfer all the results on the unit circle to the interval [-1,1]. If G is not simply connected then one has to take the Green's star domain into account, giving some difficulties in generalizing Szegö's theory to sets different from [-1,1].

Suppose the spectral measure is given by $\mu(x) = w(x)dx + \mu_2(x)$ where $w(x)$ is positive almost everywhere on [-1,1] and μ_2 is a singular measure (with respect to the Lebesgue measure) on [-1,1]. The spectral measure μ is said to belong to the *Szegö class* (on [-1,1]) if

$$\int_{-1}^{1} \frac{\log w(x)}{\sqrt{1-x^2}}\, dx > -\infty . \tag{1.3.4}$$

Clearly (1.3.4) implies that $w(x) > 0$ almost everywhere on [-1,1]. Actually, Szegö's condition gives a restriction on how close w can be to zero. Szegö's basic result is

Theorem 1.7. (Szegö [175], p. 297; Freud [61], p. 245; Geronimus [75], p. 148).

Suppose that the spectral measure μ belongs to the Szegö class on [-1,1] then

$$\lim_{n\to\infty} \frac{p_n(x;\mu)}{(x+\sqrt{x^2-1})^n} = \tag{1.3.5}$$

$$= \frac{1}{\sqrt{2\pi}} \exp\{-\frac{1}{4\pi}\int_{-\pi}^{\pi} \log[w(\cos t)|\sin t|] \frac{z+e^{-it}}{z-e^{-it}}\, dt\}$$

uniformly on closed subsets of $\bar{\mathbb{C}}\backslash[-1,1]$, where $z = x + \sqrt{x^2-1}$ and $|x+\sqrt{x^2-1}|>1$ when $x \in \bar{\mathbb{C}}\backslash[-1,1]$.

Notice that the singular part μ_2 does not enter into the asymptotic formula (1.3.5). An immediate consequence is the asymptotic behaviour of the leading coefficient

$$\lim_{n\to\infty} k_n/2^n = \frac{1}{\sqrt{\pi}} \exp\{-\frac{1}{2\pi}\int_{-1}^{1} \frac{\log w(t)}{\sqrt{1-t^2}}\, dt\}$$

which follows by setting $x = \infty$ in (1.3.5). Excellent treatments of Szegö's theory can be found in Szegö [175], Freud [61] and Geronimus [75]. Theorem 1.7 can be restated in a more convenient form (for our purpose) :

<u>Lemma 1.8.</u> Suppose that the spectral measure belongs to the Szegö class, then uniformly on closed subsets of $\bar{\mathbb{C}}\backslash[-1,1]$

$$\lim_{n\to\infty} \frac{p_n(x;\mu)}{(x+\sqrt{x^2-1})^n} = \frac{1}{\sqrt{2\pi}} \exp\left\{-\frac{\sqrt{x^2-1}}{2\pi}\int_{-1}^{1} \frac{\log[w(y)\sqrt{1-y^2}]}{\sqrt{1-y^2}} \frac{dy}{x-y}\right\} \tag{1.3.6}$$

and

$$\lim_{n\to\infty}\left\{\frac{p_n'(x;\mu)}{p_n(x;\mu)} - \frac{n}{\sqrt{x^2-1}}\right\} = \frac{1}{\sqrt{x^2-1}}\frac{1}{2\pi}\int_{-1}^{1} \frac{\log[w(y)\sqrt{1-y^2}]}{\sqrt{1-y^2}} \frac{xy-1}{(x-y)^2}\, dy . \tag{1.3.7}$$

<u>Proof</u> : Let $g(t) = \log[w(\cos t)|\sin t|]$ and $M(z)$ be the integral on the right hand side of (1.3.5), then

$$M(z) = \int_{-\pi}^{\pi} g(t) \frac{z+e^{-it}}{z-e^{-it}}\, dt = \left(\int_{-\pi}^{0} + \int_{0}^{\pi}\right) g(t) \frac{z+e^{-it}}{z-e^{-it}}\, dt$$

$$= 2\int_0^{\pi} g(t) \frac{z-\frac{1}{z}}{z-2\cos t+\frac{1}{z}}\, dt .$$

Clearly $z+\frac{1}{z} = 2x$ and $z-\frac{1}{z} = 2\sqrt{x^2-1}$, so that

$$M(z) = 2\sqrt{x^2-1}\int_0^{\pi} \frac{g(t)}{x-\cos t}\, dt .$$

If we set $y = \cos t$ then (1.3.6) follows from (1.3.5). A well-known result in the theory of analytic functions tells us that both sides of (1.3.6) may be differentiated (Rudin [167], Thm. 10.28) giving

$$\lim_{n\to\infty} \frac{p_n(x;\mu)}{(x+\sqrt{x^2-1})^n}\left\{\frac{p_n'(x;\mu)}{p_n(x;\mu)} - \frac{n}{\sqrt{x^2-1}}\right\} = \frac{1}{\sqrt{2\pi}} e^{h(x)}h'(x)$$

uniformly on closed subsets of $\bar{\mathbb{C}}\setminus[-1,1]$, where the definition of h is obvious. From (1.3.6) we then obtain

$$\lim_{n\to\infty}\left\{\frac{p_n'(x;\mu)}{p_n(x,\mu)} - \frac{n}{\sqrt{x^2-1}}\right\} = h'(x) ,$$

and if one works out the derivative of h one finds (1.3.7). ∎

We will now show how Szegö's result leads to a second order result for the asymptotic distribution of the zeros of orthogonal polynomials. In order to do so, we need to introduce a new notion. Let μ and λ be two (complex) measures on the real line, then we define for every $f \in C^\infty$ with compact support

$$\text{(1.3.8)} \qquad \mu\Delta\lambda(f) = \iint \frac{f(x)-f(y)}{x-y}\, d\mu(x)d\lambda(y) .$$

In the appendix we will show that this defines a (Schwarz) distribution of order one or zero, meaning that (1.3.8) also makes sense when f and f' are merely bounded. We call $\mu\Delta\lambda$ the *Stieltjes convolution* of μ and λ. If we denote the *Stieltjes transform* of the measure μ by

$$\text{(1.3.9)} \qquad S(\mu;z) = \int_{-\infty}^{\infty} \frac{d\mu(x)}{z-x}$$

then the main property of $\mu\Delta\lambda$ is that

$$\text{(1.3.10)} \qquad S(\mu\Delta\lambda;z) = S(\mu;z)S(\lambda;z)$$

where the Stieltjes transform of a distribution Λ is given by

$$S(\Lambda;z) = \Lambda f_z \quad ; \quad f_z(x) = \frac{1}{z-x} .$$

Theorem 1.9. (Van Assche-Teugels [196]). Suppose that the spectral measure μ belongs to the Szegö class and that $\log w$ is of bounded variation on every closed interval in $(-1,1)$. Define λ by

$$\lambda(A) = \frac{1}{2\pi} \int_A \sqrt{1-y^2}\, d[\log(w(y)\sqrt{1-y^2})] \quad , \quad A \subset [-1,1] \ . \tag{1.3.11}$$

If λ is of bounded variation on $[-1,1]$ then

$$\lim_{n\to\infty} S(n(\nu_n - \mu_E);x) = S(\mu_E\Delta\lambda;x) \tag{1.3.12}$$

uniformly on closed subsets of $\bar{\mathbb{C}}\setminus[-1,1]$, where μ_E is the arcsin measure given by (1.3.1).

<u>Proof</u> : From (1.2.9) and the integral

$$\frac{1}{\pi}\int_{-1}^{1} \frac{1}{\sqrt{1-x^2}} \frac{dx}{z-x} = \frac{1}{\sqrt{z^2-1}} \qquad z \in \mathbb{C}\setminus[-1,1] \tag{1.3.13}$$

we find that

$$S(n(\nu_n - \mu_E);x) = \frac{p_n'(x;\mu)}{p(x;\mu)} - \frac{n}{\sqrt{x^2-1}}$$

and we can therefore use Lemma 1.8. Let

$$h(y) = \frac{1}{\sqrt{1-y^2}} \log[w(y)\sqrt{1-y^2}]$$

then some elementary calculus gives

$$\int_{-1}^{1} \frac{\sqrt{1-y^2}}{x-y}\, d[\sqrt{1-y^2}\, h(y)] = \frac{1-y^2}{x-y} h(y)\Bigg|_{-1}^{1} + \int_{-1}^{1} h(y) \frac{xy-1}{(x-y)^2}\, dy \ .$$

The integrated term is zero because by the definition of h

$$(1-y)h(y) = 2\pi\sqrt{1-y^2} \int_0^y \frac{1}{\sqrt{1-t^2}}\, d\lambda(t) + 2\pi\sqrt{1-y^2}\log w(0) \ .$$

The second term on the right hand side vanishes as y tends to 1, the first term becomes zero since

$$\lim_{y\to 1-} \int_0^1 \frac{\sqrt{1-y^2}}{\sqrt{1-t^2}} I_{[0,y]}(t)\, d\lambda(t) = 0$$

which follows by using Lebesgue's theorem and the bounded variation of λ. A similar reasoning holds for $y \to -1$. Consequently

$$\lim_{n\to\infty} S(n(\nu_n - \mu_E);x) = \frac{1}{\sqrt{x^2-1}} \frac{1}{2\pi} \int_{-1}^{1} \frac{\sqrt{1-y^2}}{x-y}\, d \log [w(y) \sqrt{1-y^2}] .$$

The right hand side is the product of two Stieltjes transforms, which together with (1.3.10) gives the desired result. ∎

This theorem indicates that the sequence $\{n(\nu_n - \mu_E) : n = 1,2,\ldots\}$ converges in some sense to $\mu_E \Delta\lambda$. This convergence however is not weak since the total variation of $n(\nu_n - \mu_E)$ is equal to $2n$ which tends to infinity, meaning that $\{n(\nu_n - \mu_E) : n = 1,2,\ldots\}$ is not of uniform bounded variation (compare this with Thm. A.1 in the appendix). However, we can prove the following result.

Theorem 1.10. (Van Assche-Teugels [196]; Nevai [142], Thm. 9). Suppose that the spectral measure μ satisfies the condition in Thm. 1.9. If f is an analytic function in some open set D that contains $[-1,1]$, then

(1.3.14) $$\lim_{n\to\infty} n\left\{\frac{1}{n}\sum_{j=1}^{n} f(x_{j,n}) - \frac{1}{\pi}\int_{-1}^{1} \frac{f(y)}{\sqrt{1-y^2}}\,dy\right\}$$

$$= \frac{1}{2\pi^2} \int_{-1}^{1} \int_{-1}^{1} \frac{\sqrt{1-t^2}}{\sqrt{1-x^2}}\, \frac{f(x)-f(t)}{x-t}\, dx\, d[\log(w(t)\sqrt{1-t^2})] .$$

Proof : Let δ be the distance between $[-1,1]$ and $\mathbb{C}\setminus D$, then $\delta > 0$ so that we can define the closed curve

$$\Gamma = \{z \in \mathbb{C} : \rho(z,[-1,1]) = \tfrac{\delta}{2}\}$$

where $\rho(z,[-1,1])$ is the distance from z to $[-1,1]$. Clearly $\Gamma \subset D$ and if we set $K_n = n(\nu_n - \mu_E)$ then Cauchy's theorem gives

$$\int_{-1}^{1} f(y)\,dK_n(y) = \frac{1}{2\pi i} \int_{-1}^{1} \int_{\Gamma} \frac{f(z)}{z-y}\,dz\,dK_n(y) .$$

Fubini's theorem yields

$$\int_{-1}^{1} f(y)\,dK_n(y) = \frac{1}{2\pi i} \int_{\Gamma} f(z) S(K_n;z)\,dz .$$

The curve Γ is compact, hence we can use (1.3.12) to find

$$\lim_{n\to\infty}\int_{-1}^{1} f(y)dK_n(y) = \frac{1}{2\pi i}\int_{\Gamma} f(z)S(\mu_E\Delta\lambda;z)dz .$$

Another application of Fubini's theorem gives the result. ∎

As an example we deal with Jacobi polynomials. In this case there is no singular part for the spectral measure and the weight function is given by (0.1.4). This weight function clearly belongs to the Szegö class. It turns out that for this case we are able to impose weaker conditions on the function f in Theorem 1.10. :

Theorem 1.11. (Van Assche-Teugels [196]). For the Jacobi polynomials $\{P_n^{(\alpha,\beta)}(x) : n = 1,2,\ldots\}$ we have

$$\mu_E\Delta\lambda = \frac{\alpha+\beta+1}{2}\mu_E - \frac{2\alpha+1}{4}\delta_1 - \frac{2\beta+1}{4}\delta_{-1} \tag{1.3.15}$$

where δ_a is a degenerate probability measure with all its mass concentrated at the point a. Moreover we have for every differentiable function f for which $f' \in L^1$

$$\lim_{n\to\infty} n\left\{\frac{1}{n}\sum_{j=1}^{n} f(x_{j,n}) - \frac{1}{\pi}\int_{-1}^{1}\frac{f(y)}{\sqrt{1-x^2}}dy\right\} \tag{1.3.16}$$

$$= \frac{\alpha+\beta+1}{2}\int_{-1}^{1}\frac{f(y)}{\sqrt{1-y^2}}dy - \frac{2\alpha+1}{4}f(1) - \frac{2\beta+1}{4}f(-1) .$$

Proof : The measure λ defined in (1.3.11) becomes

$$\lambda(A) = \frac{1}{2\pi}\int_A \frac{(\beta-\alpha)-(\alpha+\beta+1)y}{\sqrt{1-y^2}}dy \qquad A \subset [-1,1]$$

so that

$$S(\mu_E\Delta\lambda;x) = \frac{1}{2\pi}\frac{1}{\sqrt{x^2-1}}\int_{-1}^{1}\frac{(\beta-\alpha)-(\alpha+\beta+1)y}{\sqrt{1-y^2}}\frac{dy}{x-y} .$$

Some elementary calculus gives

$$S(\mu_E\Delta\lambda;x) = -\frac{2\alpha+1}{4}\frac{1}{x-1} - \frac{2\beta+1}{4}\frac{1}{x+1} + \frac{\alpha+\beta+1}{2}\frac{1}{\sqrt{x^2-1}}$$

from which the expression (1.3.15) follows. Let K_n be the distribution function of $n(\nu_n - \mu_E)$, then K_n is the difference of a continuous distribution function and a function that makes jumps of size 1 at the zeros of p_n, therefore the supremum of K_n is attained at one of the zeros of p_n (say $x_{j,n}$) and is equal to $K_n(x_{j,n}+)$ or $K_n(x_{j,n}-)$. This means that

$$\|K_n\|_\infty \leq 1 + \max_{1 \leq j \leq n} |K_n(x_{j,n})| .$$

For Jacobi polynomials we have

$$\frac{n}{\pi} \operatorname{Arccos}(-x_{j,n}) = j + O(1)$$

uniformly for $j = 1,2,\ldots,n$ and $n = 1,2,3,\ldots$ (Szegö [175]; Thm. 8.9.1) which implies that $\{\|K_n\|_\infty : n = 1,2,3,\ldots\}$ is bounded. Integration by parts gives

$$\int_{-1}^{1} f(x)dK_n(x) = - \int_{-1}^{1} f'(x)K_n(x)dx$$

from which

$$\left|\int_{-1}^{1} f(x)dK_n(x)\right| \leq \|K_n\|_\infty \|f'\|_1$$

easily follows. The operators $\{\Lambda_n : n = 1,2,\ldots\}$ defined by

$$\Lambda_n f = \int_{-1}^{1} f(x)dK_n(x) = n\left\{\frac{1}{n} \sum_{j=1}^{n} f(x_{j,n}) - \frac{1}{\pi} \int_{-1}^{1} f(x) \frac{dx}{\sqrt{1-x^2}}\right\}$$

are therefore bounded in $\{f : \|f'\|_1 < \infty\}$ and since (1.3.16) already holds for polynomials on account of Theorem 1.10 we can use standard continuity arguments (Banach-Steinhaus theorem) to find the result.

■

Similar results, obtained by other methods, can be found in Nevai [142].

We can use all these results about the distribution of the zeros of orthogonal polynomials to construct quadrature formulas with simple weights. Suppose we want to approximate the integral

$$I(f) = \int_{-1}^{1} f(x)dx$$

using a quadrature formula with nodes at the zeros of orthogonal polynomials. Of course the optimal choice would be to use the Gauss-Jacobi quadrature (Lemma 0.2) but then one needs to compute the Christoffel numbers. To avoid this we will use weights that depend on the zeros in a simple way. A first suggestion is

$$(1.3.17)\qquad I_n(f) = \pi \int_{-1}^{1} f(x)\sqrt{1-x^2}\, d\nu_n(x) = \frac{\pi}{n} \sum_{j=1}^{n} f(x_{j,n})\sqrt{1-x_{j,n}^2}\ .$$

If $f(x)\sqrt{1-x^2}$ is continuous on [-1,1] and if we use zeros of orthogonal polynomials for which the spectral measure satisfies the conditions given in Theorem 1.2 (with $E = [-1,1]$) then $I_n(f)$ will converge to $I(f)$. This result even holds if $f(x)\sqrt{1-x^2}$ is merely Riemann integrable on [-1,1]. In order to have an idea of the rate of convergence we suppose that $f(x)\sqrt{1-x^2}$ is analytic in an open set that contains [-1,1], because then we find from Theorem 1.10 that

$$\lim_{n\to\infty} n\{I_n(f) - I(f)\} = \pi \int_{-1}^{1} \sqrt{1-x^2}\, f(x)\, d\mu_E\Delta\lambda(x)$$

from as soon as the spectral measure satisfies the conditions given there. This shows that the quadrature formula is $O(\frac{1}{n})$. An alternative formula is

$$(1.3.18)\qquad I_n^*(f) = \frac{2n}{\alpha+\beta+1+2n}\{I_n(f) + \frac{2\alpha+1}{4} f^*(1) + \frac{2\beta+1}{4} f^*(-1)\}$$

where $f^*(x) = \pi f(x)\sqrt{1-x^2}$ and the zeros of Jacobi polynomials are used. Theorem 1.11 then yields

$$\lim_{n\to\infty} n\{I_n^*(f) - I(f)\} = 0$$

whenever f' is absolutely integrable on [-1,1]. This quadrature formula is therefore better and has the order $o(\frac{1}{n})$. In order to emphasize the difference between I_n and I_n^* we compare these quadrature formulas with the Gauss-Jacobi quadrature for Chebyshev polynomials (Abramowitz-Stegun [1], formulas 25.4.38 and 25.4.40). Let $f(x) = (1-x^2)^{-1/2}g(x)$, with g continuous on [-1,1] then for $\alpha = \beta = -1/2$

$$I_n^*(f) = I_n(f) = \frac{\pi}{n}\sum_{j=1}^{n} g(x_{j,n}) \quad , \quad x_{j,n} = -\cos(\frac{2j-1}{2n}\pi)$$

which is exactly the Gauss-Jacobi formula for the Chebyshev polynomials of the first kind. If we let $f(x) = \sqrt{1-x^2}\, h(x)$, with h continuous on [-1,1] and $\alpha = \beta = 1/2$, then

$$I_n^*(f) = \frac{n}{n+1} I_n(f) = \frac{\pi}{n+1}\sum_{j=1} h(x_{j,n})(1-x_{j,n}^2) \quad , \quad x_{j,n} = -\cos\frac{j\pi}{n+1}$$

which is the Gauss-Jacobi formula for the Chebyshev polynomials of the second kind. Notice that the definition of I_n^* suggests that in order to have a good quadrature formula one should include the points ± 1 with appropriate weights.

1.3.2. Beyond the Szegö class

A natural question next is how the asymptotic behaviour of orthogonal polynomials on [-1,1] changes when the spectral measure does not belong to the Szegö class. A typical example are the Pollaczek polynomials. Szegö's condition (1.3.4) is violated because the weight function (0.1.22) goes to zero exponentially fast near the end-points $\pm$ 1. The appropriate asymptotic behaviour for these polynomials is given by

Theorem 1.12. (Szegö [175], appendix; Nevai [138], p. 82). For the Pollaczek polynomials with $\lambda = 1/2$ we have

$$(1.3.19)\qquad \lim_{n\to\infty} \frac{p_n(x)}{(x+\sqrt{x^2-1})^n}\, n^{-\frac{ax+b}{2\sqrt{x^2-1}}}$$

$$= \frac{\left(2\sqrt{x^2-1}\,(x-\sqrt{x^2-1})\right)^{-\frac{1}{2}+\frac{ax+b}{2\sqrt{x^2-1}}}}{\Gamma\left(\frac{1}{2}+\frac{ax+b}{2\sqrt{x^2-1}}\right)}$$

uniformly on closed sets of $\bar{\mathbb{C}}\setminus[-1,1]$, where the square root is such that $|x+\sqrt{x^2-1}| > 1$ for $x \in \bar{\mathbb{C}}\setminus[-1,1]$.

Suppose that we have a system of orthogonal polynomials with a spectral measure that looks very much like the spectral measure for Pollaczek polynomials. One hopes that the asymptotic behaviour of the polynomials then also resembles the asymptotic behaviour of the Pollaczek polynomials. More generally, suppose we know the asymptotic behaviour of a sequence $\{p_n(x;\mu_0) : n = 0,1,2,\ldots\}$ of orthogonal polynomials with spectral measure μ_0 on [-1,1] (which is not necessarily in the Szegö class). If μ is a measure that is very much like the measure μ_0, then how does $p_n(x;\mu)$ relate to $p_n(x;\mu_0)$ when n tends to infinity ? This question has been dealt with by Máté-Nevai-Totik in a sequence of papers ([121],[122],[126],[127],[150]). The basic result is

Theorem 1.13. (Máté-Nevai-Totik [126]). Let μ_0 be probability measure on [-1,1] such that $d\mu_0/dx$ is positive almost everywhere on [-1,1]. Suppose that the spectral measure μ is absolutely continuous with respect to μ_0, with Radon-Nikodym derivative $g = d\mu/d\mu_0$ and that there exists a polynomial R such that Rg and Rg^{-1} belong to $L_\infty(\mu_0)$. If $z = x + \sqrt{x^2-1}$, with $|z| > 1$ for $x \in \mathbb{C}\setminus[-1,1]$, then

$$(1.3.20)\qquad \lim_{n\to\infty} \frac{p_n(x;\mu)}{p_n(x;\mu_0)} =$$

$$\exp\left\{-\frac{1}{4\pi}\int_{-\pi}^{\pi}\log g(\cos t)\,\frac{z+e^{-it}}{z-e^{-it}}\,dt\right\}$$

holds uniformly on compact subsets of $\mathbb{C}\setminus[-1,1]$. Moreover

$$\lim_{n\to\infty}\frac{k_n(\mu)}{k_n(\mu_0)} = \exp\left\{-\frac{1}{2\pi}\int_{-1}^{1}\frac{\log g(t)}{\sqrt{1-t^2}}\,dt\right\}. \tag{1.3.21}$$

The proof of this theorem uses some deep results for orthogonal polynomials on the unit circle. Actually Szegö's result (Theorem 1.7) corresponds in a way to the case where the comparison measure μ_0 is the arcsin measure (1.3.1), for which the orthogonal polynomials are the Chebyshev polynomials of the first kind.

Let us now investigate how the second order behaviour of the zeros of orthogonal polynomials is affected when one deals with a spectral measure that does not belong to the Szegö class :

Theorem 1.14. Let μ_0 be the spectral measure for Pollaczek polynomials with $\lambda = 1/2$ and let μ absolutely continuous with respect to μ_0 with Radon-Nikodym derivative $g = d\mu/d\mu_0$. If there exists a polynomial R such that $Rg^{\pm 1} \in L_\infty(\mu_0)$ then

$$\lim_{n\to\infty} S\left(\frac{n}{\log n}(\nu_n - \mu_E);x\right) = -\frac{1}{4}\frac{1}{\sqrt{x^2-1}}\left(\frac{b-a}{x+1}+\frac{b+a}{x-1}\right) \tag{1.3.22}$$

and if f is analytic in an open set that contains [-1,1], then

$$\lim_{n\to\infty}\frac{n}{\log n}\left\{\frac{1}{n}\sum_{j=1}^{n} f(x_{j,n}) - \frac{1}{\pi}\int_{-1}^{1}\frac{f(x)}{\sqrt{1-x^2}}\,dx\right\} \tag{1.3.23}$$

$$= \frac{a-b}{4\pi}\int_{-1}^{1}\frac{f(t)-f(-1)}{t+1}\,\frac{dt}{\sqrt{1-t^2}} - \frac{a+b}{4\pi}\int_{-1}^{1}\frac{f(t)-f(1)}{t+1}\,\frac{dt}{\sqrt{1-t^2}}.$$

Proof : As in Lemma 1.8 we are allowed to take the derivative on both side of (1.3.19). This yields, uniformly on closed sets of $\bar{\mathbb{C}}\setminus[-1,1]$

$$\lim_{n\to\infty}\frac{n}{\log n}\left\{\frac{1}{n}\frac{p_n'(x;\mu_0)}{p_n(x;\mu_0)} - \frac{1}{\sqrt{x^2-1}}\right\} = -\frac{1}{4}\frac{1}{\sqrt{x^2-1}}\left(\frac{b-a}{x+1}+\frac{b+a}{x-1}\right).$$

If we do the same thing with (1.3.20) then

$$\lim_{n\to\infty}\left\{\frac{p_n'(x;\mu)}{p_n(x;\mu)} - \frac{p_n'(x;\mu_0)}{p_n(x;\mu_0)}\right\} = h(x)$$

where h is some function, that can be determined explicitely. Combine these two formulas to find (1.3.22). The result in (1.3.23) follows in exactly the same way as in Theorem 1.10. ∎

Notice that the rate of convergence of ν_n to μ_E in this case is $O(\frac{\log n}{n})$, while in Theorem 1.10 we had $O(\frac{1}{n})$.

1.4. Orthogonal polynomials on Julia sets

Let T be a monic polynomial of degree $k \geq 2$,

$$T(z) = z^k + a_{k-1}z^{k-1} + \ldots + a_0 \tag{1.4.1}$$

and let T_n be its nth iterate :

$$T_0(z) = z \quad , \quad T_n(z) = T_{n-1}(T(z)) \ .$$

Clearly T_n is a polynomial of degree k^n. Define the set J as the closure of all repulsive fixpoints, i.e. those $z \in \mathbb{C}$ for which $T_n(z) = z$ and $|T_n'(z)| > 1$. This set is the *Julia set* for the polynomial T, named after the French mathematician *Gaston Julia* (1893-1978) who made an intense study of this set [92]. The Julia set J is also the set of points $z \in \mathbb{C}$ for which $\{T_n(z) : n = 1,2,3,\ldots\}$ is not a normal family. Define the domain of attraction of infinity as

$$A(\infty) = \{z \in \bar{\mathbb{C}} : \lim_{n \to \infty} T_n(z) = \infty\}$$

then J is also equal to the boundary of $A(\infty)$. Julia sets can be defined for rational functions in a similar way and play an important role in the theory of iterations of rational functions. This theory was developped in the beginning of this century by Pierre Fatou [57] and Gaston Julia [92]. A survey of their results can be found in a paper by Brolin [32]. We always assume that T is a polynomial as given by (1.4.1). Some properties of the Julia set J are

Lemma 1.15. (Brolin [32]). If J is the Julia set for the polynomial T given in (1.4.1), then J is compact, non-empty and completely invariant, i.e.

$$T(J) = J = T^{-1}(J) \ . \tag{1.4.2}$$

The capacity of J is one and $\text{supp}(\mu_J) = J$.

Pitcher and Kinney [155] were the first to point out the relation between the iterations of T and orthogonal polynomials on J. Assume that the Julia set J is real and let $\{\hat{p}_n(x;\mu_J) : n = 0,1,2,\ldots\}$ be the monic orthogonal polynomials with spectral measure μ_J equal to the equilibrium measure on J, then

$$(1.4.3)\qquad \begin{cases} \hat{p}_1(x;\mu_J) = x + a_{k-1}/k\ , & \\ \hat{p}_{nk}(x;\mu_J) = \hat{p}_n(T(x);\mu_J)\ , & n = 0,1,2,\ldots \\ \hat{p}_{k^n}(x;\mu_J) = T_n(x) + a_{k-1}/k\ , & n = 0,1,2,\ldots \end{cases}$$

(Barnsley-Geronimo-Harrington [12]; Bessis-Moussa [27]; Pitcher-Kinney [155]). This result remains valid when J is a complex set. If we modify the orthogonality relation as

$$(1.4.4)\qquad \int_J p_n(x;\mu_J)\overline{p_m(x;\mu_J)}d\mu_J(x) = \delta_{m,n}\ .$$

For a real Julia set J we have by Theorem 1.2

$$(1.4.5)\qquad \lim_{n\to\infty} |p_n(z;\mu_J)|^{1/n} = \exp\left\{\int_J \log|z-x|d\mu_J(x)\right\}$$

(recall that C(J) = 1) and this convergence is uniform on every compact subset of $\mathbb{C}\setminus(X_1 \cup X_2)$, with X_1 the accumulation points of the zeros of $\{p_n(x;\mu_J) : n = 1,2,\ldots\}$ and X_2 the zeros that occur infinitely often. If we denote the convex hull of J by co(J) then the convergence will hold uniformly on compact sets of $\mathbb{C}\setminus co(J)$. Together with (1.4.3) this implies

$$(1.4.6)\qquad \lim_{n\to\infty} |T_n(z)|^{1/k^n} = \exp\left\{\int_J \log|z-x|d\mu_J(x)\right\} = F(z)\ .$$

For F we have

$$(1.4.7)\qquad F(T(z)) = \exp\left\{\int_J \log|T(z)-x|d\mu_J(x)\right\}$$

$$= \exp\left\{\int_J \log \prod_{j=1}^{k} |z - T^{-1}_{(j)}(x)|d\mu_J(x)\right\}$$

where $\{T^{-1}_{(j)} : j = 1,2,\ldots,k\}$ are the inverse functions of T. The equilibrium measure can be shown to be *invariant* under T, meaning

$$(1.4.8)\qquad \mu_J(T^{-1}(A)) = \mu_J(A)$$

for every Borel set $A \subset J$. More is true since

$$\mu_J(T^{-1}_{(j)}(A)) = \frac{1}{k}\,\mu_J(A) \qquad j = 1,2,\ldots,k, \tag{1.4.9}$$

(this property tells that μ_J is *balanced* under T). If we use this property in (1.4.7) then we obtain

$$F(T(z)) = \{F(z)\}^k \qquad z \in \mathbb{C}\backslash J \tag{1.4.10}$$

which is an interesting functional equation for F.

If $T(z) = z^2 - c$ with $c \geq 2$, then the Julia set will be real. If $c = 2$ then $J = [-2,2]$ and we are back to the theory developped in § 1.3. If $c > 2$ then J is real and totally disconnected. The Lebesgue measure of J is zero and $J \subset [-a,a]$ with $a = \frac{1}{2} + \sqrt{\frac{1}{4} + c}$. The Julia set J is a *Cantor set* that is constructed in the following way (Moussa [136]). In figure 1.1 we have plotted the polynomial $T(z) = z^2 - c$ $(c > 2)$.

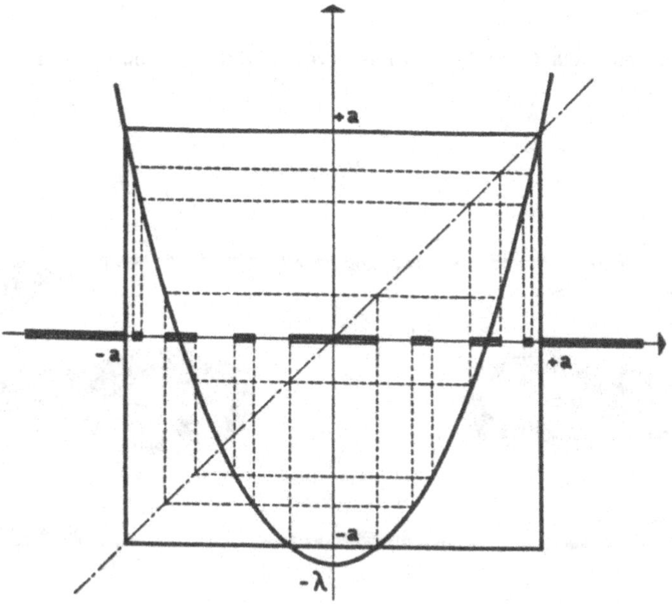

Fig. 1.1 : construction of the Julia set of $z^2 - \lambda$ $(\lambda > 2)$

Construct the square with vertices at $(\pm a, \pm a)$, then the first bisector and the polynomial will intersect at (a,a), since a is a *fixpoint* of the polynomial T. If z belongs to (a,∞) or $(-\infty,-a)$ then $\{T_n(z) : n = 1,2,\ldots\}$ will tend to infinity and the same will be true if one of the inverse images of z under T belongs to (a,∞) or $(-\infty,-a)$. The Julia set can therefore be constructed by removing all the inverse images of (a,∞) and $(-\infty,-a)$ from the real axis. This construction is very similar with the construction of the classical Cantor set.

Notice that Theorem 1.2 also applies to other measures than μ_J. If the spectral measure μ is equal to $\mu_1 + \mu_2$, with μ_1 absolutely continuous and μ_2 singular with respect to μ_J and if the accumulation points of $\mathrm{supp}(\mu_2)$ are in J, then

$$\lim_{n\to\infty} |p_n(x;\mu)|^{1/n} = F(z)$$

whenever $d\mu_1/d\mu_J$ is positive almost everywhere with respect to μ_J. In figure 1.2 we have plotted the equipotential curves for the potential associated with the Julia set for $z^2 - 3$. The limit of $|p_n(x)|^{1/n}$ will always be the same on such an equipotential curve. Theorem 1.2 also tells us that, under the appropriate conditions for the spectral measure

$$\lim_{n\to\infty} k_n^{-1/n} = 1 = C(J) .$$

We will give an upperbound for $1/k_n$ which gives an idea of how k_n behaves as $n \to \infty$.

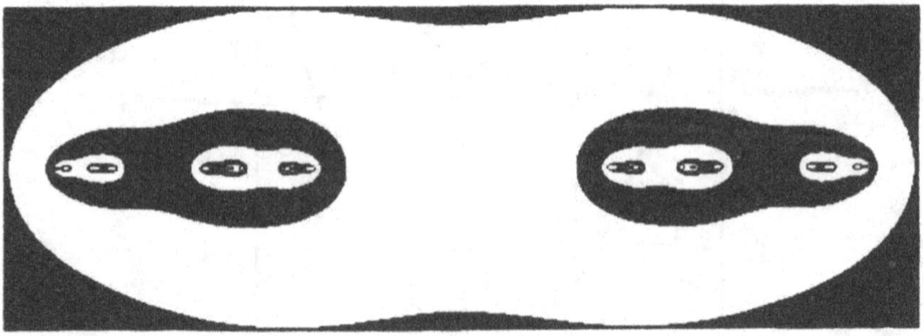

fig. 1.2 : equipotential lines for the Julia set of $z^2 - 3$.

Theorem 1.16. (Van Assche [190]) Let J be the Julia set for a polynomial T given by (1.4.1) (J is allowed to be complex) and μ be a probability measure on J, then there exists a constant $K \geq 1$ such that

$$\frac{1}{k_n^2} = \int_J |\hat{p}_n(x;\mu)|^2 d\mu(x) \leq K^m \quad , \quad k^m \leq n < k^{m+1} \tag{1.4.11}$$
$$m = 0,1,2,\ldots$$

If J is complex we use the orthogonality (1.4.4) .

Proof : We use induction on m. Let

$$K = \max_{0 \leq j < k} \sup_{x \in J} |\hat{p}_j(x;\mu_J)|^2 \tag{1.4.12}$$

then $K \geq 1$ (since $\hat{p}_0(x;\mu_J) = 1$) and $K < \infty$ (since J is compact). Use the minimal property (Lemma 0.5) (this property holds also in the complex plane if one uses the orthogonality (1.4.4)) then for $0 \leq n < k$

$$\int_J |\hat{p}_n(x;\mu)|^2 d\mu(x) \leq \int_J |\hat{p}_n(x;\mu_J)|^2 d\mu(x) \leq K$$

so that (1.4.11) holds for $m = 0$. Suppose next that (1.4.11) holds up to $m - 1$ and let n be an integer such that $k^m \leq n < k^{m+1}$. Determine j $(0 < j < k)$ such that $jk^m \leq n < (j+1)k^m$, then the minimal property yields

$$\int_J |\hat{p}_n(x;\mu)|^2 d\mu(x) \leq \int_J |\hat{p}_{jk^m}(x;\mu_J)|^2 |\hat{p}_{n-jk^m}(x;\mu)|^2 d\mu(x) .$$

From (1.4.3) we obtain that $\hat{p}_{jk^m}(x;\mu_J) = \hat{p}_j(T_m(x);\mu_J)$ and since $T(J) = J$ we have

$$\int_J |\hat{p}_n(x;\mu)|^2 d\mu(x) \leq K \int_J |\hat{p}_{n-jk^m}(x;\mu)|^2 d\mu(x) .$$

Clearly $n - jk^m < k^m$ so that we can use the induction hypothesis, which gives the desired result.

∎

A special case of this bound was given for the polynomials $\{\hat{p}_n(x;\mu_J) : n = 0,1,2\ldots\}$ with $T(z) = z^2 - c$ $(c > 2)$ by Barnsley-Geronimo-Harrington [14].

We now have a closer look at the family $\{T(z) = z^2 - c : c > 2\}$. This family of polynomials and their iterations have been studied by Barnsley-Geronimo-Harrington [13] - [15] and Bessis-Mehta-Moussa [26]. From (0.2.6) we know that the monic orthogonal polynomials satisfy a recurrence relation

$$\hat{p}_{n+1}(x;\mu_J) = x\hat{p}_n(x;\mu_J) - R_n^0\hat{p}_{n-1}(x;\mu_J) \qquad n = 0,1,2,\ldots \tag{1.4.14}$$

with $R_n^0 > 0$ $(n = 1,2,\ldots)$. The other recurrence coefficients are zero because $T(z) = z^2 - c$ is symmetric around zero. Barnsley-Geronimo-Harrington [15] and Bessis-Moussa [27] have shown that

$$\begin{cases} R_{2n}^0 R_{2n-1}^0 = R_n^0 \\ R_{2n}^0 + R_{2n+1}^0 = c \end{cases} \qquad R_0^0 = 0 \tag{1.4.15}$$

so that the polynomials $\{\hat{p}_n(x;\mu_J) : n = 0,1,2,\ldots\}$ can be constructed recursively. The recurrence relations (1.4.15) for the coefficients $\{R_n^0 : n = 1,2,\ldots\}$ are not linear, which makes things a lot more difficult. However, one can show that for $c > 2$

$$\lim_{n\to\infty} R^0_{m2^n+s} = R_s^0 \tag{1.4.16}$$

and the convergence is uniform in m and s when $c > 3$ (this last condition can be weakened). We will use this fact to study the asymptotic behaviour of orthogonal polynomials on a Julia set.

<u>Theorem 1.17.</u> Suppose that a sequence of orthogonal polynomials satisfies the recurrence relation

$$\hat{p}_{n+1}(x) = (x - A_n)\hat{p}_n(x) - R_n\hat{p}_{n-1}(x) \qquad n = 0,1,2,\ldots \tag{1.4.17}$$

with $\hat{p}_0(x) = 1$, $\hat{p}_{-1}(x) = 0$, $A_n \in \mathbb{R}$ and $R_{n+1} > 0$ $(n = 0,1,2,\ldots)$, with

$$\lim_{n\to\infty} |R_n - R_n^0| = 0 \quad , \quad \lim_{n\to\infty} A_n = 0 \tag{1.4.18}$$

where $\{R_n^0 : n = 0,1,2,\ldots\}$ is given by (1.4.15), then

$$\lim_{n\to\infty} \frac{\hat{p}_{m2^n+s}(x)}{\hat{p}_{m2^n}(x)} = \hat{p}_s(x;\mu_J) \tag{1.4.19}$$

uniformly on compact sets of $\mathbb{C}\setminus(X_1 \cup X_2)$, where X_1 are the accumulation points of the zeros of $\{\hat{p}_{m2^n}(x) : n = 0,1,2,\ldots\}$ and X_2 are those zeros that occur infinitely often.

Proof : The result is immediate for $s = 0$. Suppose $s = 1$, then from the recurrence relation (1.4.17) we find

$$(1.4.20) \qquad \frac{\hat{p}_{m2^n+1}(x)}{\hat{p}_{m2^n}(x)} = x - A_{m2^n} - R_{m2^n} \frac{\hat{p}_{m2^n-1}(x)}{\hat{p}_{m2^n}(x)}$$

By (0.2.10) and (0.2.11) we have

$$\frac{\hat{p}_{m2^n-1}(x)}{\hat{p}_{m2^n}(x)} = \sum_{j=1}^{m2^n} \frac{a_j}{x - x_j}$$

where $\{x_j : j = 1,2,\ldots,m2^n\}$ are the zeros of $\hat{p}_{m2^n}(x)$ and $\{a_j : j = 1,2,\ldots,m2^n\}$ are positive numbers with

$$\sum_{j=1}^{m2^n} a_j = 1 .$$

If K is a compact set in $\mathbb{C}\backslash(X_1 \cup X_2)$ then K contains at most a finite number of zeros of $\{p_{m2^n}(x) : n = 1,2,\ldots\}$ and each of these zeros is a zero of at most a finite number of polynomials. This means that there exists an integer N such that $K \cap (X_1 \cup X_2 \cup Z_N)$ is empty, where Z_N are the zeros of $\{p_{m2^n}(x) : n > N\}$. If δ is the distance between K and $X_1 \cup X_2 \cup Z_N$ then δ is strictly positive because K is compact and $X_1 \cup X_2 \cup Z_N$ is closed and

$$\left|\frac{\hat{p}_{m2^n-1}(x)}{\hat{p}_{m2^n}(x)}\right| < \sum_{j=1}^{m2^n} \frac{a_j}{|x - x_j|} < \frac{1}{\delta} .$$

Therefore $\{\hat{p}_{m2^n-1}(x)/\hat{p}_{m2^n}(x) : n > N\}$ is uniformly bounded on K. If we use (1.4.18) and (1.4.16) then

$$\lim_{n \to \infty} R_{m2^n} = 0 \quad , \quad \lim_{n \to \infty} A_{m2^n} = 0$$

so that (1.4.20) implies

$$\lim_{n \to \infty} \frac{\hat{p}_{m2^n+1}(x)}{\hat{p}_{m2^n}(x)} = x = \hat{p}_1(x;\mu_J) ,$$

uniformly on compact subsets of $\mathbb{C}\setminus(X_1 \cup X_2)$ meaning that (1.4.19) also holds for $s = 1$. Suppose next that (1.4.19) holds for s-2 and s-1, then the recurrence relation (1.4.17) shows that

$$\frac{\hat{p}_{m2^n+s}(x)}{\hat{p}_{m2^n}(x)} = (x - A_{m2^n+s-1}) \frac{\hat{p}_{m2^n+s-1}(x)}{\hat{p}_{m2^n}(x)} - R_{m2^n+s-1} \frac{p_{m2^n+s-2}(x)}{\hat{p}_{m2^n}(x)} .$$

Clearly by (1.4.18) and (1.4.16)

$$\lim_{n\to\infty} R_{m2^n+s-1} = R^0_{s-1} \; ; \quad \lim_{n\to\infty} A_{m2^n+s-1} = 0$$

so that

$$\lim_{n\to\infty} \frac{\hat{p}_{m2^n+s}(x)}{\hat{p}_{m2^n}(x)} = x\hat{p}_{s-1}(x;\mu_J) - R^0_{s-1}\hat{p}_{s-2}(x;\mu_J) = \hat{p}_s(x;\mu_J)$$

uniformly on compact sets of $\mathbb{C}\setminus(X_1 \cup X_2)$; which is what we wanted to prove. ∎

As a special case we find

$$\lim_{n\to\infty} \frac{\hat{p}_{m2^n+s}(x;\mu_J)}{\hat{p}_{m2^n}(x;\mu_J)} = \hat{p}_s(x;\mu_J)$$

and this is true uniformly on compact subsets of $\mathbb{C}\setminus J$ since the zeros of $\{\hat{p}_{m2^n}(x;\mu_J) = \hat{p}_m(T_n(x);\mu_J) : n = 1,2,\ldots\}$ are the inverse images of the zeros of $\hat{p}_m(x;\mu_J)$ and these accumulate on J. This result means that for large n the polynomial $\hat{p}_{m2^n+s}(x;\mu_J)$ becomes the product of $\hat{p}_{m2^n}(x;\mu_J)$ and $\hat{p}_s(x;\mu_J)$. More general results of this kind are given by Bessis-Geronimo-Moussa [24].

CHAPTER 2 : ASYMPTOTICALLY PERIODIC RECURRENCE COEFFICIENTS

In this chapter we start from the recurrence relation

$$xp_n(x) = a_{n+1}p_{n+1}(x) + b_np_n(x) + a_np_{n-1}(x) \qquad n = 0,1,2,...$$

$$p_{-1}(x) = 0 \quad , \quad p_0(x) = 1$$

where $a_{n+1} > 0$ and $b_n \in \mathbb{R}$ $(n = 0,1,2,...)$. In recent years there has been a lot of effort to relate the recurrence coefficients $\{a_{n+1}, b_n : n = 0,1,2,...\}$ to the spectral measure μ for the orthogonal polynomials $\{p_n(x) : n = 0,1,2,...\}$ (e.g. Askey-Ismail [8]; Case [34]; Chihara [39] - [44]; Dombrowski-Nevai [50]; Geronimo [69]; Máté-Nevai-Totik [117],[123],[128]; Nevai [138],[141],[143]). We suppose that the recurrence coefficients are asymptotically periodic with a period $N \geq 1$. This problem is old and certain aspects of it (in terms of periodic continued fractions) were considered by Stieltjes [174], Perron [153] and Geronimus [78],[79]. Most of the results in this Chapter have appeared in [74].

2.1. Periodic recurrence coefficients

In this section we consider the recurrence relation

$$a^o_{n+1}q_{n+1}(x) + b^o_nq_n(x) + a^o_nq_{n-1}(x) = xq_n(x) \qquad n = 0,1,2,... \tag{2.1.1}$$

with recurrence coefficients that are periodic with period N

$$\begin{cases} a^o_{n+N+1} = a^o_{n+1} \\ b^o_{n+N} = b^o_n \end{cases} \qquad n = 0,1,2,... \tag{2.1.2}$$

We will always take $a^o_o = a^o_N$. The solution of this recurrence relation with initial conditions

$$q_0(x) = 1 \quad , \quad q_{-1}(x) = 0 \tag{2.1.3}$$

consists of a sequence of polynomials $\{q_n(x) : n = 0,1,2,...\}$ with positive leading coefficient and by Favard's theorem (Lemma 0.3) there exists a probability measure μ_0 such that

$$\int_{-\infty}^{\infty} q_n(x)q_m(x)d\mu_0(x) = \delta_{m,n} \qquad m,n \geq 0 . \tag{2.1.4}$$

We also introduce the *kth associated polynomials* $\{q_n^{(k)}(x) : n = 0,1,2,\ldots\}$ which are the solution of the recurrence relation

$$a^0_{n+k+1}q^{(k)}_{n+1}(x) + b^0_{n+k}q^{(k)}_n(x) + a^0_{n+k}q^{(k)}_{n-1}(x) = xq_n(x) \tag{2.1.5}$$

with initial conditions

$$q^{(k)}_{-1}(x) = 0 \quad , \quad q^{(k)}_0(x) = 1 . \tag{2.1.6}$$

If $\{q_{n,1}(x) : n = 0,1,2,\ldots\}$ and $\{q_{n,2}(x) : n = 0,1,2,\ldots\}$ are two solutions of (2.1.1) then their *Wronskian* $W(q_{n,1},q_{n,2})$ is defined as

$$W(q_{n,1},q_{n,2}) = a^0_{n+1}\{q_{n+1,1}(x)q_{n,2}(x) - q_{n,1}(x)q_{n+1,2}(x)\} . \tag{2.1.7}$$

One can easily verify that this Wronskian is independent of n. From the general theory of linear recurrence relations it follows that two solutions $\{q_{n,1}(x) : n = 0,1,2,\ldots\}$ and $\{q_{n,2}(x) : n = 0,1,2,\ldots\}$ are linearly independent if and only if $W(q_{n,1},q_{n,2}) \neq 0$. The associated polynomials $\{q^{(1)}_{n-1}(x) : n = 0,1,2,\ldots\}$ are a solution of (2.1.1) with initial conditions

$$q^{(1)}_{-1}(x) = 1 \quad , \quad q^{(1)}_{-2}(x) = -\frac{a^0_1}{a^0_0} \tag{2.1.8}$$

and the Wronskian $W(q_n,q^{(1)}_{n-1})$ is given by

$$W(q_n,q^{(1)}_{n-1}) = -a^0_1 \tag{2.1.9}$$

which is different from zero, meaning that $\{q_n(x) : n = 0,1,2,\ldots\}$ and $\{q^{(1)}_{n-1}(x) : n = 0,1,2,\ldots\}$ are linearly independent.

The periodicity of the recurrence coefficients implies that there is a recurrence relation for q_{n+2N}, q_{n+N} and q_n :

<u>Lemma 2.1.</u> Let $\{q_{n,1}(x) : n = 0,1,2,\ldots\}$ be a solution of (2.1.1), then

$$q_{n+2N,1}(x) = \left\{q_N(x) - \frac{a^0_N}{a^0_{N+1}} q^{(1)}_{N-2}(x)\right\} q_{n+N,1}(x) - q_{n,1}(x) \qquad n = 0,1,2,\ldots \tag{2.1.10}$$

Proof : From the periodicity of the recurrence coefficients it follows that $\{q_{n+N}(x) : n = 0,1,2,...\}$ and $\{q^{(1)}_{n+N-1}(x) : n = 0,1,2,...\}$ are solutions of the recurrence relation (2.1.1), so that

$$q_{n+N}(x) = Aq_n(x) + Bq^{(1)}_{n-1}(x) \tag{2.1.11}$$

$$q^{(1)}_{n+N-1}(x) = Cq_n(x) + Dq^{(1)}_{n-1}(x) \tag{2.1.12}$$

with A,B,C,D independent of n. If we let n = 0 and n = -1 we find

$$\left\{\begin{array}{ll} A = q_N(x) \quad ; & B = -\dfrac{a^o_N}{a^o_{N+1}}\, q_{N-1}(x) \; ; \\ C = q^{(1)}_{N-1}(x) \quad ; & D = -\dfrac{a^o_N}{a^o_{N+1}}\, q^{(1)}_{N-2}(x) \; . \end{array}\right. \tag{2.1.13}$$

Change n to n + N in (2.1.11) and eliminate $q^{(1)}_{n+N-1}$ using (2.1.12) and $q^{(1)}_{n-1}(x)$ using (2.1.5), then

$$q_{n+2N}(x) = (A + D)q_{n+N}(x) + (BC - AD)q_n(x) \; .$$

By means of (2.1.9) we find that BC - AD = -1 which shows that (2.1.10) is valid for $\{q_n(x) : n = 0,1,2,...\}$. In a similar way one can show that (2.1.10) is also valid for $\{q^{(1)}_{n-1}(x) : n = 0,1,2,...\}$. This proves the result since every solution of (2.1.1) is a linear combination of these two solutions. ∎

Corollary : If the recurrence coefficients satisfy (2.1.2) then for every k > 0

$$q^{(k)}_N(x) - \frac{a^o_{N+k}}{a^o_{N+k+1}}\, q^{(k+1)}_{N-2}(x) = q_N(x) - \frac{a^o_N}{a^o_{N+1}}\, q^{(1)}_{N-2}(x) \; . \tag{2.1.14}$$

Proof : The polynomials $\{q^{(k)}_n(x) : n = 0,1,2,...\}$ satisfy a recurrence relation with periodic coefficients and therefore by Lemma 2.1

$$q^{(k)}_{n+2N}(x) = \left\{q^{(k)}_N(x) - \frac{a^o_{N+k}}{a^o_{N+k+1}}\, q^{(k+1)}_{N-2}(x)\right\} q^{(k)}_{n+N}(x) - q^{(k)}_n(x) \qquad n = 0,1,2,...$$

On the other hand $\{q^{(k)}_{n-k}(x) : n = 0,1,2,...\}$ satisfies (2.1.1) so that Lemma 2.1 (with n changed to n+k) implies that $\{q^{(k)}_n(x) : n = 0,1,2,...\}$ also satisfies (2.1.10). A comparison of both recurrence relations then gives the desired result. ∎

Equation (2.1.10) is a recurrence relation of order 2N with coefficients independent of n. The characteristic equation for this recurrence relation is

$$\alpha^{2N} - \left\{q_N(x) - \frac{a_N^o}{a_{N+1}^o} q_{N-2}^{(1)}(x)\right\}\alpha^N + 1 = 0 .$$

One solution of this characteristic equation is $\alpha_1^N = \omega^N$, where

$$\omega^N(x) = \frac{1}{2}\left\{q_N(x) - \frac{a_N^o}{a_{N+1}^o} q_{N-2}^{(1)}(x) + \rho(x)\right\} \tag{2.1.15}$$

where

$$\rho(x) = \left\{\left(q_N(x) - \frac{a_N^o}{a_{N+1}^o} q_{N-2}^{(1)}(x)\right)^2 - 4\right\}^{1/2} \tag{2.1.16}$$

with the square root chosen such that

$$\lim_{z \to \infty} \frac{\rho(z)}{z^N} = k_N = \prod_{j=1}^{N} \frac{1}{a_j^o} > 0 .$$

Another solution of the characteristic equation is $\alpha_2^N = \omega^{-N}$, where

$$\omega^{-N}(x) = \frac{1}{2}\left\{q_N(x) - \frac{a_N^o}{a_{N+1}^o} q_{N-2}^{(1)}(x) - \rho(x)\right\} . \tag{2.1.17}$$

For a closer examination of the function $\{\rho(x)\}^2$ we introduce

$$Q_N^{\pm}(x) = q_N(x) - \frac{a_N^o}{a_{N+1}^o} q_{N-2}^{(1)}(x) \pm 2 . \tag{2.1.18}$$

Denote the zeros of $Q_N^{\pm}$ in increasing order by $\{x_i^{\pm} : i = 1,2,\ldots,N\}$, the zeros of q_{N-1} by $\{x_{j,N-1} : j = 1,\ldots,N-1\}$ and those of $q_{N-1}^{(1)}$ by $\{x_{j,N-1}^{(1)} : j = 1,\ldots,N-1\}$.

<u>Lemma 2.2.</u> : (Geronimus [79]; Kac and Van Moerbeke [93],[198]). The zeros of $\{\rho(x)\}^2$ are real (but not necessarily simple) and

$$x_N^- > x_N^+ > x_{N-1,N-1}, x_{N-1,N-1}^{(1)} > x_{N-1}^+ > x_{N-1}^- > x_{N-2,N-1}, x_{N-2,N-1}^{(1)} \tag{2.1.19}$$

$$\ldots > x_{1,N-1}, x_{1,N-1}^{(1)} > \begin{cases} x_1^- > x_1^+ & N \text{ odd} \\ x_1^+ > x_1^- & N \text{ even} . \end{cases}$$

Furthermore Q_N^+ or Q_N^- will have a double zero if and only if $|q_N(x_{i,N-1})| = 1$ and this double zero is equal to $x_{i,N-1}$.

Proof : From (2.1.9) we find

$$a_N^o q_N(x_{j,N-1}) q_{N-2}^{(1)}(x_{j,N-1}) = -a_{N+1}^o$$

so that

$$Q_N^{\pm}(x_{j,N-1}) = q_N(x_{j,N-1}) + \frac{1}{q_N(x_{j,N-1})} \pm 2 \ . \tag{2.1.20}$$

The sign of $q_N(x_{j,N-1})$ is equal to $(-1)^{N-j}$ (since the zeros of q_N and q_{N-1} interlace). As $|x+\frac{1}{x}| > 2$ $(x \in \mathbb{R})$ this implies that $Q_N^{\pm}$ changes sign N-1 times which shows that the zeros of $\{\rho(x)\}^2$ are real and that the zeros of q_{N-1} interlace with the zeros of $Q_N^{\pm}$. The zeros of $q_{N-1}^{(1)}$ and q_N also interlace so that by the same argument one can show that the zeros of $q_{N-1}^{(1)}$ and Q_N^+ interlace. To find (2.1.19) we notice that, for large x, $Q_N^{\pm}$ are positive and therefore $x_N^+ < x_N^-$. At $x_{N-1,N-1}$ we see that $Q_N^{\pm}$ is negative meaning that x_N^+ and x_N^- are both to the right of $x_{N-1,N-1}$. At $x_{N-2,N-1}$ $Q_N^{\pm}$ is positive meaning that x_{N-1}^+ and x_{N-1}^- are to the left of $x_{N-1,N-1}$ and to the right of $x_{N-2,N-1}$ and since $Q_N^+ - Q_N^- = 4$ it follows that $x_{N-1}^+ > x_{N-1}^-$. If we continue like this we find (2.1.19). A zero of Q_N^+ or Q_N^- can only be a double zero at a zero $x_{i,N-1}$ and by (2.1.20) this can only happen if $|q_N(x_{i,N-1})| = 1$. Notice that in this case $x_{i,N-1} = x_{i,N-1}^{(1)}$. ∎

Define the set E by

$$E = [x_N^+, x_N^-] \cup [x_{N-1}^-, x_{N-1}^+] \cup \ldots \cup \begin{cases} [x_1^+, x_1^-] & N \text{ odd} \\ [x_1^-, x_1^+] & N \text{ even} \end{cases} \tag{2.1.21}$$

so that E exist of at most N disjoint intervals. Let F be the set

$$F = (x_{N-1}^+, x_N^+) \cup (x_{N-2}^-, x_{N-1}^-) \cup \ldots \cup \begin{cases} (x_1^-, x_2^-) & N \text{ odd} \\ (x_1^+, x_2^+) & N \text{ even} \end{cases} \tag{2.1.22}$$

which consists of the open intervals between the intervals of E. Also define the polynomial

$$U_{N-1}(x) = Q_N^+(x)' = Q_N^-(x)' \ .$$

These definitions and Lemma 2.2 imply that each interval in F contains one zero of

$q_{N-1}, q_{N-1}^{(1)}$ and U_{N-1}. Figure 2.1 shows the construction of the sets E and F.

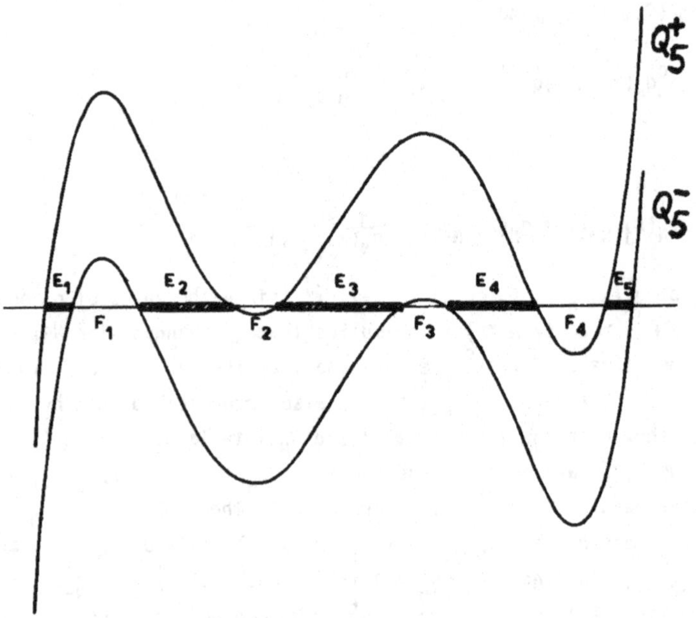

fig. 2.1 : construction of the sets E and F.

The set E consists of a finite number of intervals so that it is not difficult to construct the Green's function g_E. Clearly

$$\lim_{z \to \infty} \frac{|\omega(z)|}{|z|} = k_N^{1/N}$$

and since the product $Q_N^+ Q_N^-$ is negative on E we also have $|\omega(z)| = 1$ whenever $z \in E$. Therefore

$$g_E(z) = \log|\omega(z)|$$

and the capacity of E is

$$C(E) = k_N^{-1/N} = \left(\prod_{j=1}^{N} a_j^o\right)^{1/N} > 0 .$$

If one chooses the Nth root appropriately then ω will map the region $G = \bar{\mathbb{C}}\backslash E$ into the exterior of the unit circle. However ω will not be a one-to-one mapping since

G is not simply connected. Notice that

$$(2.1.23) \qquad \omega'(z) = \frac{\omega(z)U_{N-1}(z)}{N\rho(z)}$$

which means that $\omega'(z) = 0$ at the zeros of U_{N-1} and in the neighborhood of these zeros ω will not be invertible. Actually the Green's star domain S for G is $S = \bar{\mathbb{C}}\setminus(E \cup F)$, with E and F given by (2.1.21) and (2.1.22), and ω is a conformal mapping that maps S to the exterior of the unit circle minus radial segments emanating from the roots of unity (see figure 2.2).

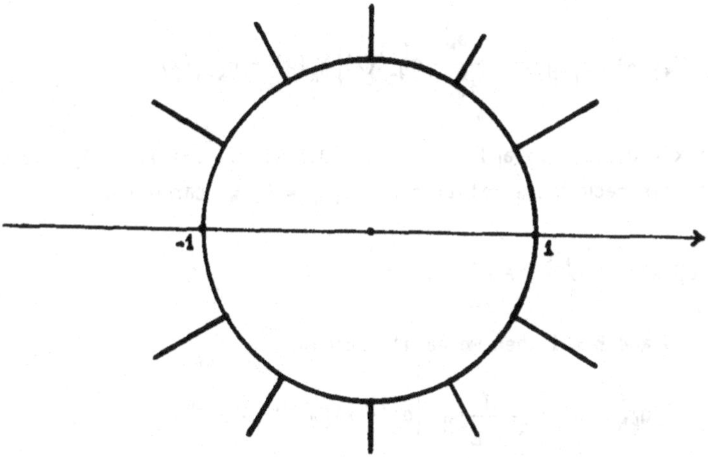

fig. 2.2 : the image of $\mathbb{C}$ under ω : the circle is the image of E, the radial segments are the images of F.

If we define

$$R(z) = \left\{q_N(z) - \frac{a_N^o}{a_{N+1}^o} q_{N-2}^{(1)}(z)\right\}^2 - 4$$

then $\rho(z) = \sqrt{R(z)}$ $(z \in \mathbb{C}\setminus E)$ and $\sqrt{R(x \pm io)} = \pm i\sqrt{-R(x)}$ $(x \in E)$. Let K be the two sheeted Riemann surface with cuts along the intervals of E and branch points at the ends of these intervals, then K is of genus at most N-1 and $G = \bar{\mathbb{C}}\setminus E$ is one sheet of K.

The conformal mapping ω plays an important role if we want to describe the asymptotic behaviour of orthogonal polynomials with asymptotically periodic recurrence coefficients. A fundamental upperbound is given by

Lemma 2.3. There exists a positive constant A such that for every integer i

$$(2.1.24) \qquad |\omega^{-n} q_n^{(i)}(x)| < A \frac{n+N}{N + |1-\omega^{-2N}|(n+N)} \qquad x \in \bar{\mathbb{C}} .$$

Proof : Let $q_k^*(x) = q_{kN+s}^{(i)}(x)$, with i, N and s fixed, then it follows from Lemma 2.1 and its corollary that $\{q_k^* : k = 0,1,2,\ldots\}$ satisfies the recurrence

$$q_{k+1}^*(x) = \left\{ q_N(x) - \frac{a_N^o}{a_{N+1}^o} q_{N-2}^{(1)}(x) \right\} q_k^*(x) - q_{k-1}^*(x) .$$

Since $\{\omega^{kN} : k = 0,1,2,\ldots\}$ and $\{\omega^{-kN} : k = 0,1,2,\ldots\}$ are linearly independent solutions of this recurrence relation (if $\rho(x) \neq 0$) we can write

$$q_k^*(x) = A_1 \omega^{kN} + A_2 \omega^{-kN} .$$

If we set k = 0 and k = 1 then we easily obtain

$$(2.1.25) \qquad \omega^{-kN} q_{kN+s}^{(i)}(x) = \frac{1}{1-\omega^{-2N}} \left\{ q_s^{(i)}(x)(\omega^{-2kN} - \omega^{-2N}) + \omega^{-N} q_{N+s}^{(i)}(x)(1-\omega^{-2kN}) \right\} .$$

Define

$$(2.1.26) \qquad C_1 = \max_{0 \le i < N} \ \max_{0 \le s \le 2N-1} \ \sup_{x \in \mathbb{C}} |\omega^{-s} q_s^{(i)}(x)|$$

then C_1 is finite because $\lim_{x \to \infty} x/\omega(x)$ is finite, and from (2.1.25) one finds

$$|\omega^{-(kN+s)} q_{kN+s}^{(i)}(x)| \le \frac{C_1}{|1-\omega^{-2N}|} \left\{ |\omega^{-2kN} - \omega^{-2N}| + |1-\omega^{-2kN}| \right\} .$$

Use the bound

$$\frac{|1-z^{2n}|}{|1-z^2|} \le C_2 \frac{n}{1+|1-z^2|n} \qquad (|z| \le 1)$$

to find

ERRATUM

LECTURE NOTES IN MATHEMATICS, VOL. 1265
WALTER VAN ASSCHE, ASYMPTOTICS FOR ORTHOGONAL POLYNOMIALS
ISBN 3-540-18023-0

p. 111 : line 12 : $$\lim_{n\to\infty} \frac{1}{m^{2/\alpha}} \left\{ \int_{-cm_n^{1/\alpha}}^{cm_n^{1/\alpha}} |\hat{p}_{m_n}(x)|^2 \, w(x) \, dx \right\}^{1/m_n} < \beta^2$$

line 14 : $$\lim_{n\to\infty} \frac{1}{m_n^{2/\alpha}} \left\{ \int_{-1}^{1} |\hat{p}_{m_n}(cm_n^{1/\alpha}x)|^2 \, w(cm_n^{1/\alpha}x)\,dx \right\}^{1/m_n} < \beta^2$$

line 18 : $$\limsup_{n\to\infty} \frac{1}{t_n^{1/\alpha}} |\hat{p}_{t_n}(ct_n^{1/\alpha}x)|^{1/t_n} \exp\left(- \frac{|x|^\alpha}{\lambda_\alpha}\right) < \beta \, .$$

p. 154 : The statement following line 4 is false. The weak convergence does not imply that the moments of ν_n convergence since $f(x) = |x|^M$ is not a bounded function.

Replace line 5-6-7 by

The extra condition $\frac{1}{c_n} k_{n-1}/k_n = o(\sqrt{n})$ is fulfilled if the $(2 + \varepsilon)$-moment of ν_n converges. Indeed, by taking $M = 2 + \varepsilon$ $(\varepsilon > 0)$ we then have

$$|\omega^{-(kN+s)} q^{(i)}_{kN+s}(x)| < A \frac{k+1}{1+|1-\omega^{-2N}|(k+1)}$$

with $A = 2C_1C_2$, then (2.1.24) follows. ∎

We now introduce two solutions of the recurrence relation (2.1.1) that reflect the periodicity of the recurrence coefficients. These solutions correspond to the *Floquet solutions* (or *Bloch waves*) for Matthieu's equation.

Theorem 2.4. Let

(2.1.27) $$q_n^+(x) = q_{n+N}(x) - \omega^N q_n(x) ,$$

(2.1.28) $$q_n^-(x) = q_{n+N}(x) - \omega^{-N} q_n(x) ,$$

then $\{q_n^\pm(x) : n = 0,1,2,\ldots\}$ are two solutions of (2.1.1) for which $q_n^\pm(x) = \omega^{\mp}\phi_n^\pm(x)$ with $\{\phi_n^\pm(x) : n = 0,1,2,\ldots\}$ periodic sequences with period N. These two solutions are linearly independent if and only if $q_{N-1}(x) \neq 0$ and $\omega^{2N} \neq 1$.

Proof : The sequences $\{q_{n+N}(x) : n = 0,1,2,\ldots\}$ and $\{q_n(x) : n = 0,1,2,\ldots\}$ are both solutions of (2.1.1) and therefore also each linear combination of these sequences, which shows that $\{q_n^\pm(x) : n = 0,1,2,\ldots\}$ are solutions of (2.1.1). From Lemma 2.1 we find

(2.1.29) $$q_{n+2N}(x) = (\omega^N + \omega^{-N}) q_{n+N}(x) - q_n(x) .$$

Change n to n+N in (2.1.27) and (2.1.28) and substitute this in (2.1.29) then it follows that

(2.1.30) $$q_{n+N}^\pm(x) = \omega^{\mp} q_n^\pm(x)$$

from which the periodicity follows. The Wronskian of these two solutions is

(2.1.31) $$W(q_n^+, q_n^-) = a_N^0 q_{N-1}(x)(\omega^{-N} - \omega^N)$$

which proves the result. ∎

The most important properties of these solutions are

Lemma 2.5. The functions q_n^+ and q^- are analytic in $\mathbb{C}\backslash E$, real on $\mathbb{R}\backslash E$ and $q_n^+(x) = \overline{q_n^-(x)}$ where $x \in E$. Furthermore

(2.1.32) $|\omega^{-n-N} q_n^-(x)| < 2C_1 \qquad x \in \bar{\mathbb{C}}$

(2.1.33) $|\omega^{n-N+2} q_n^+(x)| < D \qquad x \in \bar{\mathbb{C}}$

where C_1 and D are positive constants.

Proof : The analytical properties follow from the definition of $q_n^{\pm}$ and the fact that ω^N and ω^{-N} are analytic in $\mathbb{C}\backslash E$, real on $\mathbb{R}\backslash E$ and $\omega^N = \overline{\omega^{-N}}$ when $x \in E$. The inequality (2.1.32) follows by setting $n = kN + s$ and using (2.1.30) and (2.1.26). In order to prove (2.1.33) we use

$$\begin{aligned} q_n^+(x)q_n^-(x) &= (q_{n+N}(x))^2 - (\omega^N + \omega^{-N})q_{n+N}(x)q_n(x) + (q_n(x))^2 \\ &= (q_{n+N}(x))^2 - q_{n+2N}(x)q_n(x) \end{aligned}$$

where Lemma 2.1 has been applied. Now $\{q_{n-m-1}^{(m+1)}(x) : n = 0,1,2,\ldots\}$ is a solution of the recurrence relation (2.1.1) and therefore can be written as a linear combination of $\{q_n(x) : n = 0,1,2,\ldots\}$ and $\{q_{n+N}(x) : n = 0,1,2,\ldots\}$, namely

$$q_{n-m-1}^{(m+1)}(x) = \frac{a_{m+1}^0}{a_N^0} \frac{1}{q_{N-1}(x)} \{q_{m+N}(x)q_n(x) - q_m(x)q_{n+N}(x)\} .$$

Change n to n + N and m to n to find

$$q_{N-1}(x)q_{N-1}^{(n+1)}(x) = \frac{a_{n+1}^0}{a_N^0} \{(q_{n+N}(x))^2 - q_{n+2N}(x)q_n(x)\}$$

so that

(2.1.34) $$q_n^+(x)q_n^-(x) = \frac{a_N^0}{a_{n+1}^0} q_{N-1}(x)q_{N-1}^{(n+1)}(x) .$$

Finally use $q_n^-(x) = O(\omega^{n+N})$ and Lemma 2.3. ∎

Lemma 2.6. The zeros of $\{q_{kN+s}^{\pm}(x) : k = 0,1,2,\ldots\}$ $(s = 0,1,\ldots,N-1)$ are real and given by the zeros of q_{N-1} and/or $q_{N-1}^{(s+1)}$. A zero $x_{j,N-1}$ of q_{N-1} will be a zero for every q_n^+ $(n = 0,1,2,\ldots)$ if and only if $|q_N(x_{j,N-1})| > 1$.

Proof : The first part follows immediately from (2.1.34). Clearly $q_{-1}^+(x_{j,N-1}) = 0$ and if $q_0^+(x_{j,N-1}) = 0$ then by the recurrence formula (2.1.1) every q_n^+ will be zero at $x_{j,N-1}$. Therefore $q_n^+(x_{j,N-1}) = 0$ for every n if and only if $q_0^+(x_{j,N-1}) = 0$.

By (2.1.27) and (2.1.15) this is possible if and only if

$$q_N(x_{j,N-1}) + \frac{a_N^o}{a_{N+1}^o} q_{N-2}^{(1)}(x_{j,N+1}) = \rho(x_{j,N-1}) .$$

Now

$$\left\{q_N(x_{j,N-1}) + \frac{a_N^o}{a_{N+1}^o} q_{N-2}^{(1)}(x_{j,N-1})\right\}^2 = (\rho(x_{j,N-1}))^2$$

and therefore $q_0^+(x_{j,N-1}) = 0$ if and only if $\rho(x_{j,N-1})$ has the same sign as

$$q_N(x_{j,N-1}) + \frac{a_N^o}{a_{N+1}^o} q_{N-2}^{(1)}(x_{j,N-1}) = q_N(x_{j,N-1}) - \frac{1}{q_N(x_{j,N-1})} .$$

The result then follows since $\rho(x_{j,N-1})$ and $q_N(x_{j,N-1})$ already have the same sign.

■

2.2. Asymptotic periodicity

We now start with a recurrence relation

(2.2.1) $$a_{n+1}p_{n+1}(x) + b_np_n(x) + a_np_{n-1}(x) = xp_n(x) \qquad n = 0,1,2,...$$

where $a_{n+1} > 0$ and $b_n \in \mathbb{R}$ $(n = 0,1,2,...)$ are such that

(2.2.2) $$\lim_{n\to\infty} |a_n - a_n^o| = 0 \quad ; \quad \lim_{n\to\infty} |b_n - b_n^o| = 0 ,$$

the sequences $\{a_n^o : n = 1,2,...\}$ and $\{b_n^o : n = 0,1,2,...\}$ are periodic and satisfy (2.1.2). The solution of (2.2.1) with initial conditions $p_0(x) = 1$ and $p_{-1}(x) = 0$ consists of orthogonal polynomials $\{p_n(x) : n = 0,1,2,...\}$ with a spectral measure μ. The kth associated polynomials are given by

(2.2.3) $$a_{n+k+1}p_{n+1}^{(k)}(x) + b_{n+k}p_n^{(k)}(x) + a_{n+k}p_{n-1}^{(k)}(x) = xp_n^{(k)}(x) \qquad n = 0,1,2,...$$

with initial conditions $p_0^{(k)}(x) = 1$ and $p_{-1}^{(k)}(x) = 0$ (when $k = 0$ we usually drop the superscript). Our aim is to compare the solutions of (2.2.1) with solutions of the periodic system (2.1.1). In order to do this we introduce two *Green's sequences* G_1 and G_2. They are the solutions of

$$(2.2.4) \qquad a^o_{n+1}G_i(x,n+1,m) + b^o_nG_i(x,n,m) + a^o_nG_i(x,n-1,m) - xG_i(x,n,m) = \delta_{n,m}$$

$$(i = 1,2)$$

with boundary conditions

$$G_1(x,n,m) = 0 \qquad\qquad n > m \quad,$$

$$G_2(x,n,m) = 0 \qquad\qquad n < m \quad.$$

They are explicitely given by

$$(2.2.5) \qquad a^o_{n+1}G_1(x,n,m) = q^{(n+1)}_{m-n-1}(x) \qquad\qquad -1 < n < m$$

$$= 0 \qquad\qquad m \leq n$$

and

$$(2.2.6) \qquad a^o_{m+1}G_2(x,n,m) = q^{(m+1)}_{n-m-1}(x) \qquad\qquad -1 < m < n$$

$$= 0 \qquad\qquad n \leq m \quad.$$

(Atkinson [10], Geronimo [71]).

<u>Theorem 2.7.</u> Let $\{p_n(x) : n = 0,1,2,...\}$ be the orthogonal polynomials that satis-
fy (2.2.1) with $p_{-1} = 0$ and $p_0 = 1$. If we define

$$\tilde{p}_n(x) = \left(\prod_{j=1}^{n} \frac{a_j}{a^o_j}\right)p_n(x)$$

then

$$(2.2.7) \qquad \tilde{p}_m(x) = q_m(x) + \sum_{n=0}^{m-1} k_1(x,n,m)\tilde{p}_n(x)$$

with

$$(2.2.8) \qquad k_1(x,n,m) = (b^o_n - b_n)G_1(x,n,m) + a^o_{n+1}\left\{1 - \left(\frac{a_{n+1}}{a^o_{n+1}}\right)^2\right\}G_1(x,n+1,m) \quad.$$

For every $x \in \bar{C}$ we have the upper bound

$$(2.2.9) \qquad |\omega^{-m}\tilde{p}_m(x)| < A \frac{m + N}{N + |1 - \omega^{-2N}|(m+N)} \exp\left\{A \sum_{n=0}^{m-1} \frac{n + N}{N + |1 - \omega^{-2N}|(n+N)}\right\}$$

where A is a positive constant and

(2.2.10) $$k_n = \frac{|b_n^o - b_n|}{a_{n+1}^o} + \frac{a_{n+1}^o}{a_{n+2}^o} \left|1 - \left(\frac{a_{n+1}}{a_{n+1}^o}\right)^2\right| .$$

Proof : Multiply (2.2.1) by $\left(\prod_{j=1}^{n} \frac{a_j}{a_j^o}\right) G_1(x,n,m)$ and (2.2.4) by $\tilde{p}_n(x)$ and substract the two resulting equations. Summation over n ranging from 0 to ∞ gives (2.2.7). To find the bound in (2.2.9) we substitute (2.2.5) into (2.2.8) and by means of Lemma 2.3 we obtain

(2.2.11) $$|\omega^{-(m-n)} k_1(x,n,m)| < A k_n \frac{m+N}{N + |1 - \omega^{-2N}|(m+N)} .$$

Write

(2.2.12) $$|\omega^{-m}| \, \tilde{p}_m(x) = \sum_{i=0}^{\infty} g_i(x,m)$$

with

$$g_0(x,m) = |\omega^{-m}| \, q_m(x)$$

and

$$g_i(x,m) = \sum_{n=0}^{m-1} |\omega^{-(m-n)}| k_1(x,n,m) g_{i-1}(x,n)$$

then from Lemma 2.3 we find

(2.2.13) $$|g_0(x,m)| < A \frac{m+N}{N + |1 - \omega^{-2N}|(m+N)} .$$

If we solve for $g_i(x,m)$ recursively then

$$g_i(x,m) = \sum_{n_1=0}^{m-1} |\omega^{-(m-n_1)}| \, k_1(x,n_1,m) \sum_{n_2=0}^{n_1-1} |\omega^{-(n_1-n_2)}| \, k_1(x,n_2,n_1)$$

$$\cdots \sum_{n_i=0}^{n_{i-1}-1} |\omega^{-(n_{i-1}-n_i)}| k_1(x,n_i,n_{i-1}) g_0(x,n_i) .$$

Use the bounds (2.2.11) and (2.2.13) to find

$$|g_i(x,m)| < A \frac{m+N}{N + |1 - \omega^{-2N}|(m+N)} \frac{1}{i!} \left\{ A \sum_{n=0}^{m-1} k_n \frac{n+N}{N + |1 - \omega^{-2N}|(n+N)} \right\}^i .$$

Insert this into (2.2.12) then (2.2.9) follows. ∎

We will now try to find two solutions $\{p_n^{\pm}(x) : n = 0,1,2,\ldots\}$ of the recurrence relation (2.2.1) that resemble the solutions $\{q_n^{\pm}(x) : n = 0,1,2,\ldots\}$ of the periodic recurrence relation as n tends to infinity. In order to do so we will impose the following conditions on the recurrence coefficients :

$$(2.2.14)\qquad \begin{cases} a_{n_0+j+1} = a^o_{n_0+j+1} \\ b_{n_0+j} = b^o_{n_0+j} \end{cases} \qquad j = 0,1,2,\ldots$$

The corresponding orthogonal polynomials will be denoted by $\{p_n(x;n_0) : n = 0,1,2,\ldots\}$ and we define $\{p_n^{\pm}(x;n_0) : n = 0,1,2,\ldots\}$ as the solutions of (2.2.1) for which

$$p_n^{\pm}(x;n_0) = q_n^{\pm}(x) \qquad\qquad n > n_0 \; .$$

It is clear that $p_n^{\pm}(x;n_0)$ and $q_n^{\pm}(x)$ indeed have the same behaviour for large enough n (actually they are equal for large enough n).

Lemma 2.8. Define

$$\tilde{p}_n^{+}(x;n_0) = \left(\prod_{j=n+1}^{\infty} \frac{a_j}{a_j^o}\right) p_n^{+}(x;n_0)$$

then

$$(2.2.15)\qquad \tilde{p}_m^{+}(x;n_0) = q_m^{+}(x) + \sum_{n=m+1}^{n_0} k_2(x,n,m)\tilde{p}_n^{+}(x;n_0)$$

where

$$(2.2.16)\qquad k_2(x,n,m) = (b_n^o - b_n)G_2(x,n,m) + a_n^o \left\{1 - \left(\frac{a_n}{a_n^o}\right)^2\right\} G_2(x,n-1,m) \; .$$

Proof : Multiply (2.2.1) by $\left(\prod_{j=n+1}^{\infty}\right)\frac{a_j}{a_j^o} G_2(x,n,m)$ and (2.2.4) by $\tilde{p}_n^{+}(x;n_0)$ and substract the resulting equations. Summation with the subscript n ranging from 0 to ∞ gives, with the appropriate boundary conditions

$$\tilde{p}_m^{+}(x;n_0) = a^o_{n_0+1}\{q^{+}_{n_0}(x)G_2(x,n_0+1,m) - q^{+}_{n_0+1}(x)G_2(x,n_0,m)\} + \sum_{n=m+1}^{n_0} k_2(x,n,m)\tilde{p}_m^{+}(x;n_0) \; .$$

The first term on the right hand side is equal to the Wronskian

$$\frac{1}{a^o_{m+1}} W(q^{(m+1)}_{n-m-1}, q^+_n) = q^+_m$$

from which (2.2.15) follows. ∎

Theorem 2.9. Let $G = \bar{\mathbb{C}}\setminus E$ and ∂G be the boundary of G on the Riemann surface K. Let

$$k'_n = \frac{|b^o_n - b_n|}{a^o_{n+1}} + \frac{a^o_n}{a^o_{n+1}} \left|1 - \left(\frac{a_n}{a^o_n}\right)^2\right|$$

and suppose that

$$(2.2.17) \qquad \sum_{n=0}^{\infty} k'_n < \infty$$

then there exists a solution $\{p^+_n(x) : n = 0,1,2,\ldots\}$ of the recurrence relation (2.2.1) such that p^+_n is analytic in G and continuous on $(G \cup \partial G)\setminus\{\omega^{2N} = 1\}$ for every n. Furthermore

$$(2.2.18) \qquad \lim_{m \to \infty} |\omega^{(k-1)N} \{p^+_m(x) - q^+_m(x)\}| = 0 \qquad (m = kN + s)$$

uniformly on closed subsets of $(G \cup \partial G)\setminus\{\omega^{2N} = 1\}$. If

$$(2.2.19) \qquad \sum_{n=0}^{\infty} nk'_n < \infty$$

then p^+_n is continuous on $G \cup \partial G$ and the convergence in (2.2.18) is uniform in $G \cup \partial G$.

Proof : Let n_0 tend to infinity in (2.2.15) then we get a discrete integral equation for $\{p^+_n(x) : n = 0,1,2,3,\ldots\}$

$$(2.2.20) \qquad \tilde{p}^+_m(x) = q^+_m(x) + \sum_{n=m+1}^{\infty} k_2(x,n,m)\tilde{p}^+_n(x) .$$

Again we write

$$(2.2.21) \qquad \omega^{N(k-1)}\tilde{p}^+_m(x) = \sum_{i=0}^{\infty} g_i(x,m)$$

where $m = kN + s$,

$$g_0(x,m) = \omega^{(k-1)N} q^+_m(x)$$

and

$$g_i(x,m) = \sum_{n=m+1}^{\infty} \omega^{m-n} k_2(x,n,m) g_{i-1}(x,n) \ .$$

From Lemma 2.5 we find, since $|\omega| > 1$ on $G \cup \partial G$

$$|g_0(x,m)| < |\omega^{s+2} g_0(x,m)| < D \ .$$

Using (2.2.16), (2.2.6) and Lemma 2.3 yields

$$|g_1(x,m)| < D \sum_{n=m+1}^{\infty} Ak_n' \frac{n+N}{N + |1-\omega^{-2N}|(n+N)} \ .$$

If we solve for $g_i(x,m)$ recursively then

$$(2.2.22) \qquad |g_i(x,m)| < D \frac{1}{i!} \left\{ A \sum_{n=m+1}^{\infty} k_n' \frac{n+N}{N + |1-\omega^{-2N}|(n+N)} \right\}^i \ .$$

Every g_i is analytic in G and continuous on $G \cup \partial G$ and therefore (2.2.17) and (2.2.22) imply that (2.2.21) converges uniformly on closed subsets of $(G \cup \partial G)\backslash \{\omega^{2N} = 1\}$. Consequently $\omega^{(k-1)N} p_m^+$ $(m = kN + s)$ is analytic in G and continuous on $(G \cup \partial G)\backslash\{\omega^{2N} = 1\}$. If (2.2.19) is fulfilled then the convergence is uniform on $G \cup \partial G$, which implies the continuity on $G \cup \partial G$. From (2.2.20) we find, using (2.2.22)

$$|\omega^{(k-1)N}\{\widetilde{p}_m^+(x) - q_m^+(x)\}| < D \sum_{n=m+1}^{\infty} |\omega^{m-n} k_2(x,n,m)|$$

$$\times \exp\left\{ A \sum_{j=m+1}^{\infty} k_j' \frac{j+N}{N + |1-\omega^{-2N}|(j+N)} \right\}$$

which gives all the remaining properties stated in this theorem. ∎

<u>Corollary</u>. If (2.2.17) holds then

$$(2.2.23) \qquad \lim_{n_0 \to \infty} |\omega^{(k-1)N}\{\widetilde{p}_m^+(x) - \hat{p}_m^+(x;n_0)\}| = 0 \qquad (m = kN + s)$$

uniformly on closed subsets of $(G \cup \partial G)\backslash\{\omega^{2N} = 1\}$. When (2.2.19) holds then (2.2.23) is valid uniformly on $G \cup \partial G$.

<u>Proof</u> : If we substract (2.2.15) from (2.2.20) then

$$\omega^{(k-1)N}\{\tilde{p}_m^+(x) - \tilde{p}_m^+(x;n_0)\} = \sum_{n=n_0+1}^{\infty} k_2(x,n,m)\omega^{(k-1)N}\tilde{p}_n^+(x)$$

$$+ \sum_{n=m+1}^{\infty} k_2(x,n,m)\omega^{(k-1)N}\{\tilde{p}_n^+(x) - \tilde{p}_n^+(x;n_0)\} .$$

Use successive approximations (as in Theorems 2.7 and 2.9) to find

$$|\omega^{(k-1)N}\{\tilde{p}_m^+(x) - \tilde{p}_m^+(x;n_0)\}| \le \exp\left\{A \sum_{j=m+1}^{\infty} k_j' \frac{j+N}{N + |1-\omega^{-2N}|(j+N)}\right\}$$

$$\times \sum_{n=n_0+1}^{\infty} k_n' \frac{n+N}{N+|1-\omega^{-2N}|(n+N)} |\omega^{(k-1)N}\tilde{p}_n^+(x)|$$

Then (2.2.23) follows since $|\omega^{(k-1)N}\tilde{p}_n^+(x)|$ is bounded on closed sets of $(G \cup \partial G)\setminus\{\omega^{2N} = 1\}$ when (2.2.17) is valid, and on $G \cup \partial G$ when (2.2.19) holds. ∎

Notice that the set ∂G is not the same as E since there is a difference in approaching E from above and from below, indicated by

$$\lim_{y \to 0+} \rho(x + iy) = i\sqrt{-R(x)}$$

$$\lim_{y \to 0+} \rho(x - iy) = -i\sqrt{-R(x)} .$$

However, we will always refer to ∂G as E for simplicity, in the assumption that this causes no confusion to the reader. The set $\{\omega^{2N} = 1\}$ contains the endpoints of the intervals in E and no other points if E consists of N disjoint intervals. Recall that $G = \bar{\mathbb{C}}\setminus E$ is one of the sheets of the Riemann surface K for ρ. If we call the other sheet G' then by analytic continuation of p_n^+ to G' we can find a solution $\{p_n^-(x) : n = 0,1,\ldots\}$ of the recurrence (2.2.1) for which

$$\lim_{m \to \infty} |\omega^{(1-k)N}\{\tilde{p}_m^-(x) - q_m^-(x)\}| = 0 \qquad (m = kN + s)$$

uniformly on closed subsets of $(G' \cup \partial G')\setminus\{\omega^{2N} = 1\}$ whenever (2.2.17) is valid. If (2.2.19) is true then the convergence is uniform on $G' \cup \partial G'$. On E we have $\omega^N = \bar{\omega}^{-N}$ and $q_n^+ = \overline{q_n^-(x)}$, therefore we have $p_n^-(x) = \overline{p_n^+(x)}$ for every $x \in E$.

<u>Lemma 2.10.</u> Suppose that (2.2.17) is satisfied then for every $x \in E\setminus\{\omega^{2N} = 1\}$

$$(2.2.24) \qquad p_n(x) = \frac{f_-(x)p_n^+(x) - f_+(x)p_n^-(x)}{a_0^0 q_{N-1}(x)(\omega^{-N} - \omega^N)}$$

where

$$(2.2.25) \qquad f_\pm(x) = W(p_n,p_n^\pm) = a_0 p_{-1}^\pm(x) .$$

<u>Proof</u> : Notice that $\{p_n^+(x) : n = 0,1,2,\ldots\}$ and $\{p_n^-(x) : n = 0,1,2,\ldots\}$ are two solutions of (2.2.1) that are continuous on $E\backslash\{\omega^{2N} = 1\}$. Clearly

$$W(p_n^-,p_n^+) = a_0^0 q_{N-1}(x)(\omega^{-N} - \omega^N)$$

so that these solutions are linearly independent on $E\backslash\{\omega^{2N} = 1\}$. Writing $p_n(x) = Ap_n^+(x) + Bp_n^-(x)$ yields

$$A = \frac{W(p_n,p_n^+)}{W(p_n^-,p_n^+)} \quad ; \quad B = - \frac{W(p_n,p_n^-)}{W(p_n^-,p_n^+)}$$

which gives the result. ∎

The functions $f_+(x)$ and $f_-(x)$ correspond to the *Jost functions* from scattering theory. Later we will see that they contain a lot of information about the spectral measure μ of the orthogonal polynomials $\{p_n(x) : n = 0,1,2,\ldots\}$.

Let us separate the zeros of p_n^+ into two sets. The first set $R_1(n)$ contains all the zeros of p_n^+ that are also zeros of p_{n-1}^+, the second set $R_2(n)$ contains all the other zeros of p_n^+. One can show that whenever (2.2.17) holds all the zeros of p_n^+ in $\mathbb{C}\backslash E$ are real and that there are no zeros on $E\backslash\{\omega^{2N} = 1\}$. The zeros of $R_2(n)$ in $\mathbb{C}\backslash E$ are simple and if x_1 and x_2 are two zeros in $R_2(n)$ which are not separated by an interval of E, then there is a zero of p_{n-1}^+ and a zero of p_{n+1}^- between x_1 and x_2. If (2.2.19) holds then p_n^+ has a finite number of zeros in $\mathbb{C}\backslash E$. (Geronimo-Van Assche [74]).

The infinite Jacobi matrix J_∞ associated with the recurrence coefficients $\{a_{n+1},b_n : n = 0,1,2,\ldots\}$ corresponds to an operator from ℓ_2 to ℓ_2, where ℓ_2 are the square summable sequences $\{\psi = (\psi_n : n = 0,1,2,\ldots) : \sum_{n=0}^\infty \psi_n^2 < \infty\}$. Then x_1 will be an eigenvalue of J_∞ if and only if there exists a sequence $\psi \in \ell_2$, different from 0; such that $J_\infty\psi = x_1\psi$. This means exactly that ψ is a solution of the recurrence relation (2.2.1) with boundary condition $\psi_{-1} = 0$ and therefore

$$\psi_n = \psi_0 p_n(x_i) \qquad n = 0,1,2,\ldots$$

Since $\psi \in \ell_2$ we have

$$\sum_{n=0}^{\infty} |p_n(x_1)|^2 < \infty$$

and this is well known to be a necessary and sufficient condition in order that the spectral measure μ has positive mass at x_1 (Chihara [39], p. 70). This means that the eigenvalues of J_∞ are equal to the discrete mass points of the spectral measure μ. Next we will show that the eigenvalues of J_∞ are essentially given by the zeros of f_+ :

Theorem 2.11. : Suppose that (2.2.17) holds. If $f_+(x_1) = 0$ and $x_1 \in R_2(-1)\backslash E$ or if $f_+(x_1) = f_+'(x_1) = 0$ with $x_1 \in R_1(-1)\backslash E$ then x_1 is an eigenvalue of J and

$$p_n(x_1) = \frac{\hat{p}_n^+(x_1)}{\hat{p}_0^+(x_1)} \tag{2.2.26}$$

with

$$\hat{p}_n^+(x) = \begin{cases} p_n^+(x) & \text{if } f_+'(x_1) \neq 0 \\ \dfrac{p_n^+(x)}{x - x_1} & \text{if } f_+'(x_1) = 0 . \end{cases} \tag{2.2.27}$$

Proof : First of all we notice that, since $\{p_n^+(x) : n = 0,1,2,\ldots\}$ satisfies the recurrence relation (2.2.1), one can show that

$$a_{n+1}\{p_{n+1}^+(x)(p_n^+(x))' - (p_{n+1}^+(x))'p_n^+(x)\} = \sum_{i=n+1}^{\infty} \{p_i^+(x)\}^2 \tag{2.2.28}$$

in a similar way as one proves the formula of Christoffel-Darboux (in addition, one uses $\lim_{n\to\infty} p_n^+(x) = \lim_{n\to\infty} (p_n^+(x))' = 0$ for $x \notin E$). Suppose first that $f_+(x_1) = 0$ with $x_1 \in R_2(-1)\backslash E$. Consider the sequence $(\psi_n = p_n^+(x_1) : n = 0,1,2,\ldots)$ then from (2.2.28) we find that $\psi \neq 0$ and $\psi \in \ell_2$. Moreover $J_\infty\psi = x_1\psi$ which means that x_1 is an eigenvalue of J_∞.

Suppose next that $f_+(x_1) = f_+'(x_1) = 0$ with $x_1 \in R_1(-1)\backslash E$, then $p_n^+(x_1) = 0$ for every n since $p_n^+(x_1)$ is a solution of the recurrence relation (2.2.1) with boundary conditions $p_{-1}^+(x_1) = 0$ and $p_0^+(x_1) = 0$. By (2.2.20) we find that $q_n^+(x_1) = 0$ for every n. Therefore we can divide $p_n^+(x)$ and $q_n^+(x)$ by $x - x_1$ for every x and n without changing (2.2.1) or (2.2.20). Since $q_{nN-1}^+(x_1) = \omega^{-nN}q_{N-1}(x_1) = 0$ we find that

$$\lim_{x\to x_1} \frac{q_{nN-1}^+(x)}{x - x_1} = \lim_{x\to x_1} \omega^{-nN} \frac{q_{-1}^+(x)}{x - x_1} = \omega^{-nN}(x_1)q_{N-1}'(x_1) \neq 0$$

(since zeros of orthogonal polynomials are always simple), therefore for n large enough

$$\lim_{x \to x_1} \frac{p^+_{nN-1}(x)}{x - x_1} \neq 0$$

and $\{\psi_n = \hat{p}^+_n(x_1) : n = 0,1,2,\ldots\}$ will be a (nonzero) eigenvector for J_∞ with eigenvalue x_1. The expression (2.2.26) follows since both sides are a solution of the recurrence relation (2.2.1) with the same initial condition at $n = 0$ and $n = -1$.

■

Let us now return to the recurrence coefficients for which (2.2.14) holds. By using the recurrence relation and the boundary conditions at ∞ one can find an expression for the product $f_+(x)f_-(x)$:

Lemma 2.13. The function $(\omega^{-N} - \omega^N)/f_+(x;n_0)$ is continuous on E. Moreover

$$(2.2.29) \qquad f_+(x;n_0)f_-(x,n_0) = a^0_0 q_{N-1}(x)\{a^0_{n_0} p_{n_0+N-1}(x;n_0)p_{n_0}(x;n_0) - a_{n_0}p_{n_0-1}(x;n_0)p_{n_0+N}(x;n_0)\} .$$

Proof : From (2.2.24) we find for $x \in E\backslash\{\omega^{2N} = 1\}$

$$\frac{\omega^{-N} - \omega^N}{f_+(x;n_0)} p_m(x;n_0) = \frac{S(x)p^+_m(x;n_0) - p^-_m(x;n_0)}{a^0_N q_{N-1}(x)}$$

where $S(x) = f_-(x;n_0)/f_+(x;n_0)$. From the definition of p^-_n and p^+_n it follows that $|S(x)| = 1$ for $x \in E\backslash\{\omega^{2N} = 1\}$. The continuity of $(\omega^{-N} - \omega^N)/f_+(x;n_0)$ then follows since $p^+_m(x;n_0)/q_{N-1}(x)$, $p^-_m(x;n_0)/q_{N-1}(x)$, $f_+(x;n_0)$ and $f_-(x;n_0)$ are continuous on E. Substitute $n = n_0$ and $n = n_0 + N$ in (2.2.24) and multiply respectively by $p^+_{n_0+N}(x;n_0)$ and $p^+_{n_0}(x;n_0)$. The difference of the obtained equations is

$$q^+_{n_0}(x) \left\{p_{n_0+N}(x;n_0) - \omega^{-N}p_{n_0}(x;n_0)\right\} = \frac{f_+(x;n_0)q^+_{n_0}(x)q^-_{n_0}(x)}{a^0_N q_{N-1}(x)}$$

where we used $p^\pm_{n_0+n}(x;n_0) = q^+_{n_0+n}(x)$. Now use (2.1.34), then

$$(2.2.30) \qquad f_+(x;n_0) = \frac{a^0_{n_0+1}\, q^+_{n_0}(x)}{q^{(n_0+1)}_{N-1}(x)} \left\{p_{n_0+N}(x;n_0) - \omega^{-N}p_{n_0}(x;n_0)\right\}$$

and a similar result holds for $f_-(x;n_0)$. Therefore

$$f_+(x;n_0)f_-(x;n_0) = \left(\frac{a^0_{n_0+1}}{q^{(n_0+1)}_{N-1}(x)}\right)^2 q^+_{n_0}(x)q^-_{n_0}(x)$$

$$\times \left\{(p_{n_0+N}(x;n_0))^2 + (p_{n_0}(x;n_0))^2 - (\omega^N + \omega^{-N})p_{n_0+N}(x;n_0)p_{n_0}(x;n_0)\right\}.$$

Since $\{p_n(x;n_0) : n > n_0\}$, $\{p_{n+N}(x;n_0) : n > n_0\}$ and $\{q^{(n_0+1)}_{n-n_0-1}(x) : n > n_0\}$ satisfy the same recurrence relation we find

$$p_{n+N}(x;n_0) = A\ p_n(x;n_0) + B\ q^{(n_0+1)}_{n-n_0-1}(x) .$$

If we set $n = n_0$ and $n = n_0 + 1$ then we are able to determine A and B. For $n = n_0 + N$ this gives

$$q^{(n_0+1)}_{N-1}(x) = \frac{(p_{n_0+N}(x;n_0))^2 - p_{n_0}(x;n_0)p_{n_0+2N}(x;n_0)}{p_{n_0+N}(x;n_0)p_{n_0+1}(x;n_0) - p_{n_0}(x;n_0)p_{n_0+N+1}(x;n_0)} .$$

Combine all this to find

$$f_+(x;n_0)f_-(x;n_0) = a^0_0 a^0_{n_0+1} q_{N-1}(x)$$

$$\times \left\{p_{n_0+N}(x;n_0)p_{n_0+1}(x;n_0) - p_{n_0}(x;n_0)p_{n_0+N+1}(x;n_0)\right\}$$

where we used (2.1.10) to connect $p_{n_0+2N}(x;n_0)$, $p_{n_0+N}(x;n_0)$ and $p_{n_0}(x;n_0)$. Now reduce the degree of $p_{n_0+1}(x;n_0)$ and $p_{n_0+N+1}(x;n_0)$ by using the recurrence relation (2.2.1), then (2.2.29) follows. ∎

2.3. Construction of the spectral measure

In this section we construct the spectral measure μ for the orthogonal polynomials $\{p_n(x) : n = 0,1,2,...\}$ that satisfy the recurrence relation (2.2.1) with asymptotically periodic recurrence coefficients. First we deal with recurrence coefficients for which (2.2.14) applies.

Theorem 2.14. : Suppose that the recurrence coefficients satisfy (2.2.14) then the spectral measure μ, with respect to which the polynomials $\{p_n(x;n_0) : n = 0,1,2,...\}$

are orthogonal, can be written as

$$(2.3.1)\qquad d\mu(x) = \sigma(x)dx + \sum_{i=1}^{M} \rho_i dU(x - x_i)$$

where

$$(2.3.2)\qquad \sigma(x) = \frac{a_N^0\, q_{N-1}(x)}{2\pi |f_+(x;n_0)|^2} \sqrt{4 - \left\{q_N(x) - \frac{a_N^0}{a_{N+1}^0}\, q_{N-2}^{(1)}(x)\right\}^2} \quad , \quad x \in E\backslash\{\omega^{2N} = 1\}$$

and $dU(x - x_i)$ is a degenerate measure with mass one at $x_i \in \mathbb{R}\backslash E$, $f_+(x_i;n_0) = 0$ and

$$(2.3.3)\qquad \rho_i = \frac{\hat{p}_0^+(x_i;n_0)}{(\hat{f}_+(x_i;n_0))'} .$$

The functions $\{\hat{p}_n^+ : n = -1,0,1,2,\ldots\}$ are defined in (2.2.27).

<u>Proof</u> : Consider the closed curve $\Gamma = \Gamma_1 \cup \Gamma_2$, where

$$\Gamma_1 = \{z \in \mathbb{C} : |z| = 1,\ z \neq e^{\pm ik\pi/N} \quad , \quad k = 1,2,\ldots,N-1\}$$

and Γ_2 is the union of 2N-2 closed curves encircling the 2N-2 images of F under $\frac{1}{\omega}$ (see figure 2.3).

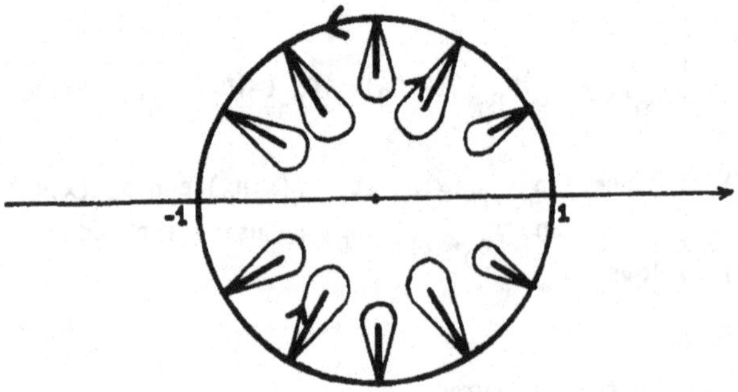

fig. 2.3

We now calculate the integral ($m < n$)

$$(2.3.4)\qquad I = -\int_\Gamma \frac{p_m(x;n_0)\,p_n^+(x;n_0)}{2\pi i\; f_+(x;n_0)}\, h'(z)dz = I_1 + I_2$$

where I_1 and I_2 are the integrals corresponding to Γ_1 and Γ_2 and h is the inverse function of $\frac{1}{\omega}$, so that $z = 1/\omega(x)$ and $x = h(z)$. Using (2.1.23) we find

$$(2.3.5)\qquad h'(z) = \frac{1}{\left.\frac{d(1/\omega)}{dx}\right|_{x=h(z)}} = - \left.\frac{N(\omega^{-N} - \omega^{N})}{\frac{1}{\omega}\left\{q_N(x) - \frac{a_N^o}{a_{N+1}^o} q_{N-2}^{(1)}(x)\right\}'}\right|_{x = h(z)}$$

and by Lemma 2.13 it therefore follows that I is well defined. Let us calculate I_1 first. Use (2.2.24) to eliminate $p_n^+(x;n_0)$ and then use $f_+(x;n_0) = \overline{f_-(x;n_0)}$ on Γ_1 to find

$$(2.3.6)\qquad I_1 = \frac{1}{2\pi i}\int_{\Gamma_1} \frac{p_m(x;n_0)p_n^-(x;n_0)}{f_-(x;n_0)} h'(z)dz$$

$$- \frac{1}{2\pi i}\int_{\Gamma_1} \frac{W(p_n^+,p_n^-)p_m(x;n_0)p_n(x;n_0)}{|f_+(x;n_0)|^2} h'(z)dz .$$

Substitute $z = e^{i\theta}$ in the first integral on the right and then change θ to $-\theta$ (in which case $p_n^-(x;n_0)$ changes to $p_n^+(x;n_0)$ and $e^{i\theta}h'(e^{i\theta})$ changes to $-e^{-i\theta}h'(e^{-i\theta})$) then we find

$$I_1 = -I_1 - \frac{1}{2\pi i}\int_{\Gamma_1} \frac{W(p_n^+,p_n^-)p_m(x;n_0)p_n(x;n_0)}{|f_+(x;n_0)|^2} h'(z)dz$$

from which one easily finds

$$I_1 = - \frac{1}{4\pi i}\int_{\Gamma_1} \frac{W(p_n^+,p_n^-)p_m(x;n_0)p_n(x;n_0)}{|f_+(x;n_0)|^2} h'(z)dz .$$

Now $\Gamma_1 = \Gamma_+ \cup \Gamma_-$ with $\Gamma_+ = \Gamma_1 \cap \{z \in \mathbb{C} : \operatorname{Im} z > 0\}$ and $\Gamma_- = \Gamma_1 \cap \{z \in \mathbb{C} : \operatorname{Im} z < 0\}$ and if we change θ to $-\theta$ in the integral over Γ_- we find

$$I_1 = - \frac{1}{2\pi i}\int_{\Gamma_+} \frac{W(p_n^+,p_n^-)p_m(x;n_0)p_n(x;n_0)}{|f_+(x;n_0)|^2} h'(z)dz .$$

If we map Γ_+ to the set E and use the fact that on Γ_+

$$W(p_n^+,p_n^-) = -i\, a_N^o\, q_{N-1}(x)\sqrt{4 - \left\{q_N(x) - \frac{a_0^o}{a_{N+1}^o} q_{N-2}^{(1)}(x)\right\}^2}$$

then we find

$$I_1 = \int_E p_m(x;n_0)p_n(x;n_0)\sigma(x)dx \tag{2.3.7}$$

with σ given by (2.3.2).

We can evaluate the integral I by using the theorem of residues. There is a possible residu for $z = 0$ and for the zeros of $f_+(x;n_0)$ that are lying in Γ. For large x we have

$$p_m(x;n_0) \sim (\prod_{j=1}^{m} \frac{1}{a_j})\, x^m$$

and from (2.1.27), (2.1.34), (2.1.28) and (2.1.17) we find

$$\frac{p_n^+(x;n_0)}{f_+(x;n_0)} \sim (\prod_{j=1}^{n} a_j)x^{-n-1}.$$

Now $z = 1/\omega(x) \sim C(E)/x$, where $C(E) = (\prod_{j=1}^{N} a_j^0)^{1/N}$ is the capacity of E, therefore

$$-\frac{p_m(x;n_0)p_n^+(x;n_0)}{f_+(x;n_0)}\, h'(z) \sim (\prod_{j=1}^{m} \frac{1}{a_j})(\prod_{i=1}^{n} a_i)(C(E))^{m-n} z^{n-m-1} \tag{2.3.8}$$

where we used $h'(z) \sim -C(E)z^{-2}$. The residu at $z = 0$ is therefore given by $\delta_{m,n}$ ($m < n$). In order to calculate the other residues we notice that all the zeros of $p_{-1}^+(x;n_0)$ that are lying in Γ belong to $R_2(-1;n_0)$. If we use (2.2.26) then

$$I = \delta_{n,m} - \sum_i \frac{p_m(h(z_i);n_0)p_n(h(z_i);n_0)p_0^+(h(z_i);n_0)}{\frac{d}{dz} f_+(h(z);n_0)\Big|_{z=z_i}}\, h'(z_i) \tag{2.3.9}$$

where the z_i are such that $f_+(h(z_i);n_0) = 0$.

Finally we calculate the integral I_2. We use the transformation $z = h(x)$ that maps the curve Γ_1 to the intervals of E (described twice) and Γ_2 to N-1 curves, each containing a component of F (see figure 2.4). Then we find

$$I_2 = -\frac{1}{2\pi i}\int_{\Gamma_2} \frac{p_m(x;n_0)p_n^+(x;n_0)}{f_+(x;n_0)}\, h'(z)dz$$

$$= -\frac{1}{2\pi i}\sum_{j=1}^{N-1}\int_{Dj} \frac{p_m(x;n_0)p_n^+(x;n_0)}{f_+(x;n_0)}\, dx$$

with D_j a curve around F_j.

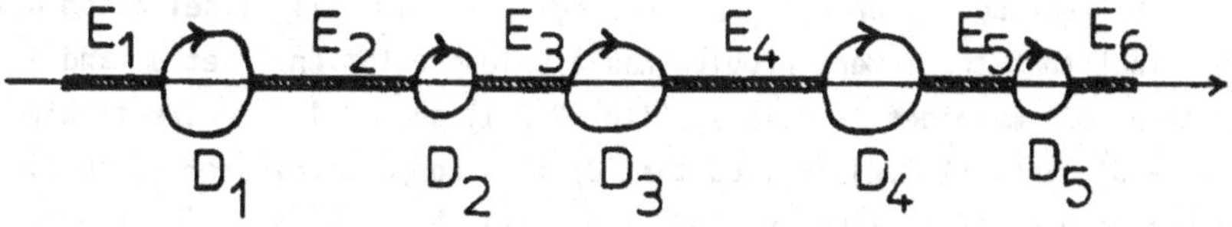

Fig. 2.4

The only zeros of $f_+(x;n_0)$ that are relevant are those in $R_2(-1;n_0)$ and those in $R_1(-1;n_0)$ for which $f_+'(x;n_0) = 0$. The theorem of residues gives

$$(2.3.10) \qquad I_2 = \sum_k p_m(x_k;n_0)p_n(x_k;n_0)\,\frac{\hat{p}_0(x_k;n_0)}{\hat{f}_+'(x_k;n_0)} \quad .$$

(the minus sign has disappeared since D_j is orientated in clock wise direction). The result then follows from (2.3.4), (2.3.7), (2.3.9) and (2.3.10). ∎

Notice that $|f_+(x;n_0)|^2$ is a polynomial, given by (2.2.29). The weight function σ has also been obtained by Geronimus [79]. If we use (2.2.24) to evaluate $p_n^+(x;n_0)$ at a zero x_k of $f_+(x;n_0)$, for which $q_{N-1}(x_k) \neq 0$, then we find

$$\rho_k = \frac{a_0^0\, q_{N-1}(x_k)\sqrt{\left\{q_N(x_k) - \frac{a_N^0}{a_{N+1}^0}\, q_{N-2}^{(1)}(x_k)\right\}^2 - 4}}{f_+'(x_k;n_0)f_-(x_k;n_0)}$$

which is exactly the formula for the mass at x_k given by Geronimus [79]. The case N=1 has also been treated by Dombrowski and Nevai [50].

Next we want to treat the asymptotically periodic case. We use a result given by Geronimus [78] :

<u>Lemma 2.15.</u> : Let μ be the spectral measure for orthogonal polynomials with recurrence coefficients that satisfy (2.2.2), then μ is the weak limit of the sequence $\{\mu_{n_0} : n_0 = 1,2,3,\ldots\}$ where μ_{n_0} is the spectral measure of the

polynomials $\{p_n(x;n_0) : n = 0,1,2,\ldots\}$. The spectral measure μ is given by $\mu = \mu_c + \mu_d$ where μ_c is a measure with a continuous distribution function with $\mathrm{supp}(\mu_c) = E$ and μ_d is a discrete measure with $\mathrm{supp}(\mu_d)$ at most denumerable and with accumulation points in E.

Proof : The weak convergence of μ_{n_0} to μ follows from Helly's selection principle and the fact that the moment problem has a unique solution. Let J_∞ and J_∞^0 be the infinite Jacobi matrices associated with (2.2.1) and (2.1.1) respectively and let $J_p = J_\infty - J_\infty^0$, then (2.2.2) implies that J_p is a compact operator (Kato [100],p. 137). A theorem of Weyl (Kato [100],p. 244) then says that the essential spectrum of J_∞ is the same as the essential spectrum of J_∞^0, which proves our result. ∎

An immediate corollary to Lemma 2.15 and Theorem 2.14 is

Theorem 2.16. Suppose that

$$\sum_{n=1}^{\infty} \left\{|b_n^0 - b_n| + |1 - (\frac{a_n}{a_n^0})^2|\right\} < \infty$$

then the spectral measure μ is given by

$$d\mu(x) = \sigma(x)dx + \sum_i \rho_i dU(x - x_i) \tag{2.3.11}$$

where

$$\sigma(x) = \frac{a_N^0\, q_{N-1}(x)}{2\pi |f_+(x)|^2} \sqrt{4 - \left\{q_N(x) - \frac{a_N^0}{a_{N+1}^0} q_{N-2}^{(1)}(x)\right\}^2} \quad , \quad x \in E\backslash\{\omega^{2N} = 1\} \tag{2.3.12}$$

and

$$\rho_i = \frac{\hat{p}_0^+(x_i)}{\hat{f}_+'(x_i)}$$

with $x_i \in \mathbb{R}\backslash E$ and $f_+(x_i) = 0$. If

$$\sum_{n=1}^{\infty} n\left\{|b_n^0 - b_n| + |1 - (\frac{a_n}{a_n^0})^2|\right\} < \infty$$

then the sum in (2.3.11) contains only a finite number of terms.

The weight function σ in (2.3.12) has an important property. Indeed, if we define the *Szegö class* for the set E to be the class of all weight functions w on E for which

$$\int_E \log w(x)\, d\mu_E(x) > -\infty ,$$

with μ_E the equilibrium measure on E, then

Theorem 2.17. Suppose that

$$\sum_{n=2}^{\infty} \log n \left\{ |b_n - b_n^o| + \left|1 - \left(\frac{a_n}{a_n^o}\right)^2\right| \right\} < \infty$$

then σ belongs to the Szegö class on E.

Proof : The equilibrium measure μ_E is given by

$$d\mu_E(x) = \frac{1}{N\pi} \frac{\left|\left\{q_N(x) - \frac{a_N^o}{a_{N+1}^o} q_{N-2}^{(1)}(x)\right\}'\right|}{\sqrt{4 - \left\{q_N(x) - \frac{a_N^o}{a_{N+1}^o} q_{N-2}^{(1)}(x)\right\}^2}} dx$$

$$= \Delta(x)dx , \qquad x \in E\backslash\{\omega^{2N} = 1\} .$$

Consider the integral

$$I = \int_E \log |f_+(x)|\, d\mu_E(x) .$$

Recall that $f_+(x) = a_o\, p_{-1}^+(x)$ and use (2.2.21) and (2.2.22) to find

$$I < C + A \int_E \sum_{n=o}^{\infty} k_n' \frac{n + N}{N + |1 - \omega^{-2N}|(n + N)} d\mu_E(x)$$

where A and C are positive constants. Now

$$\int_E f(x)d\mu_E(x) = \frac{1}{2\pi} \int_0^{2\pi} f(h(e^{i\theta}))d\theta$$

for every continuous function f (Sario-Nakai [168], Chapter III, § 2) hence

$$I < C + A \frac{1}{2\pi} \int_0^{2\pi} \sum_{n=o}^{\infty} k_n' \frac{n + N}{N + |1 - e^{2iN\theta}|\,(n + N)} d\theta .$$

Use the integral

$$\int_0^{\pi} \frac{dt}{a + b \sin t} = \frac{1}{\sqrt{b^2 - a^2}} \log \frac{b + \sqrt{b^2 - a^2}}{b - \sqrt{b^2 - a^2}} \qquad (b^2 > a^2)$$

then we can find a constant B such that

$$I < C + AB \sum_{n=0}^{\infty} \log (n + N + 1) k_n' .$$

The sum on the right hand side is finite and by (2.3.12) the result follows. ∎

2.4. Strong asymptotics

In this section we analyze the asymptotic behaviour of orthogonal polynomials with asymptotically periodic recurrence coefficients. Let us first consider the Chebyshev polynomials of the first kind (this corresponds to the case N = 1). We know that $T_n(x) = \cos n\theta$, where $x = \cos\theta$. If we put $x = \frac{1}{2}(z + \frac{1}{z})$ and $z = x + \sqrt{x^2 - 1}$ then for $z = e^{i\theta}$ one finds $x = \cos\theta$. Therefore by analytic continuation

$$T_n(x) = \frac{1}{2}(z^n + z^{-n}) \qquad z \in \mathbb{C}.$$

In analyzing the asymptotic behaviour of $T_n(z)$ one then needs to consider two separate cases : if $|z| > 1$ (which corresponds to $x \in \mathbb{C}\setminus[-1,1]$) then z^n will have the most important contribution to $T_n(x)$ and z^{-n} tends to zero if n tends to infinity, therefore

$$\lim_{n \to \infty} \frac{T_n(x)}{(x + \sqrt{x^2 - 1})^n} = 1/2 \qquad x \in \bar{\mathbb{C}}\setminus[-1,1]$$

and this convergence is uniform on closed subsets of $\bar{\mathbb{C}}\setminus[-1,1]$. If $x \in [-1,1]$ then $|z| = 1$ and both z^n and z^{-n} will contribute to $T_n(x)$. On the spectrum [-1,1] we have $T_n(x) = \cos n\theta$ which means that T_n has oscillatory behaviour on [-1,1].

We will try to use this line of thought to study the asymptotic behaviour of orthogonal polynomials with asymptotically periodic recurrence coefficients, and we will look for functions $\psi_n^+(x)$ and $\psi_n^-(x)$ that play the role of z^n and z^{-n} in the previous example.

First we derive a formula that generalizes Lemma 2.1.

Lemma 2.18. (Geronimus [79]) Let $\{p_{n,1}(x) : n = 0,1,2,\ldots\}$ be a solution of the recurrence relation (2.2.1), then

$$p_{n+2N,1}(x)\, p_{N-1}^{(n+1)}(x) = p_{n+N,1}(x)\, p_{2N-1}^{(n+1)}(x) - \frac{a_{n+1}}{a_{n+N+1}}\, p_{n,1}(x)\, p_{N-1}^{(n+N+1)}(x) \,. \tag{2.4.1}$$

Proof : First we will show that this Lemma is true for $\{p_n(x) : n = 0,1,2,\ldots\}$. Since $\{p_n(x) : n = 0,1,2,\ldots\}$, $\{p_{n-m}^{(m)}(x) : n = 0,1,2,\ldots\}$ and $\{p_{n-m-1}^{(m+1)}(x) : n = 0,1,2,\ldots\}$ are three solutions of (2.2.1) they have to be linearly dependent. One easily finds

$$p_n(x) = p_m(x)\, p_{n-m}^{(m)}(x) - \frac{a_m}{a_{m+1}}\, p_{m-1}(x)\, p_{n-m-1}^{(m+1)}(x) \,. \tag{2.4.2}$$

Change n to n + N and m to n + 1 in this formula and multiply this by $p_{2N-1}^{(n+1)}(x)$, then change n to n + 2N and m to n + 1 and multiply by $p_{N-1}^{(n+1)}(x)$. If we substract the two resulting equations, then

$$p_{n+N}(x)p_{2N-1}^{(n+1)}(x) - p_{n+2N}(x)p_{N-1}^{(n+1)}(x) = \frac{a_{n+1}}{a_{n+2}}\, p_n(x) \left\{p_{2N-2}^{(n+2)}(x)p_{N-1}^{(n+1)}(x) - p_{N-2}^{(n+2)}(x)p_{2N-1}^{(n+1)}(x)\right\} . \tag{2.4.3}$$

From (2.4.2) we also find

$$p_{2N-m}^{(n+m)}(x) = p_{N-m}^{(n+m)}(x)p_N^{(n+N)}(x) - \frac{a_{n+N}}{a_{n+N+1}}\, p_{N-m-1}^{(n+m)}(x)p_{N-1}^{(n+N+1)}(x) \,. \tag{2.4.4}$$

Let m = 1 and m = 2 in (2.4.4) and multiply this by $p_{N-2}^{(n+2)}(x)$ and $p_{N-1}^{(n+1)}(x)$ respectively, then substraction of the two resulting equations gives

$$p_{2N-2}^{(n+2)}(x)p_{N-1}^{(n+1)}(x) - p_{2N-1}^{(n+1)}(x)p_{N-2}^{(n+2)}(x) = \frac{a_{n+1}}{a_{n+N+1}}\, p_{N-1}^{(n+N+1)}(x) \tag{2.4.5}$$

where we used $W(p_{n-m-1}^{(m+1)}, p_{n-m-2}^{(m+2)}) = -a_{m+2}$. Substitute (2.4.5) into (2.4.3) then the lemma follows for $\{p_n(x) : n = 0,1,2,\ldots\}$. In a similar way one can show that the Lemma holds for $\{p_{n-1}^{(1)} : n = 0,1,2,\ldots\}$. The general result then follows since every solution of (2.2.1) is a linear combination of $p_n(x)$ and $p_{n-1}^{(1)}(x)$. ∎

Szegö's asymptotic theory for orthogonal polynomials on [-1,1] consists of studying the asymptotic behaviour of orthogonal polynomials on the unit circle and then to "translate" this to orthogonal polynomials on [-1,1]. Orthogonal polynomials on the unit circle do not satisfy a second order recurrence relation but they do satisfy a system of two first order recurrence relations (Szegö [175], p. 293). The following lemma shows how one can transform the recurrence relation (2.2.1) to a system of two recurrence relations : (see Geronimo - Case [72] for the case N = 1) :

Lemma 2.19 : A sequence of orthogonal polynomials $\{p_n(x) : n = 0,1,2,...\}$, for which the recurrence coefficients satisfy (2.2.2), can be constructed recursively by

$$(2.4.6) \qquad p_{n+N}(x) = \frac{a_{n+1}^o}{a_{n+1}} \frac{p_{N-1}^{(n+1)}(x)}{q_{N-1}^{(n+1)}(x)} \left\{(\omega^{-N} - B_n(x))p_n(x) + \omega^{\pm N}\psi_n^{\pm}(x)\right\}$$

$$(2.4.7) \qquad \psi_{n+N}^{\pm}(x) = \frac{a_{n+1}^o}{a_{n+1}} \frac{p_{N-1}^{(n+1)}(x)}{q_{N-1}^{(n+1)}(x)} \left\{\omega^{\pm N}\psi_n^{\pm}(x) + \left[\left(1 - \left(\frac{a_{n+1}}{a_{n+1}^o} \frac{q_{N-1}^{(n+1)}(x)}{p_{N-1}^{(n+1)}(x)}\right)^2\right)\omega^{\mp N} - B_n(x)\right] p_n(x)\right\}$$

where

$$(2.4.8) \qquad B_n(x) = \omega^N + \omega^{-N} - \frac{a_{n+1}}{a_{n+1}^o} q_{N-1}^{(n+1)}(x) \left\{\frac{p_N^{(n)}(x)}{p_{N-1}^{(n+1)}(x)} - \frac{a_n}{a_{n+1}} \frac{p_{N-2}^{(n-N+1)}(x)}{p_{N-1}^{(n-N+1)}(x)}\right\}$$

and $\psi_0^{\pm}(x) = p_0(x) = 1$.

Proof : Solve (2.4.6) for $\psi_n^{\pm}(x)$ and insert the result in (2.4.7) then (2.4.1) follows provided we substitute $p_{2N-1}^{(n+1)}$ there using (2.4.4) (with m = 1). ∎

If we substract (2.4.7) from (2.4.6) then

$$(2.4.9) \qquad \psi_{n+N}^{\pm}(x) = p_{n+N}(x) - \frac{a_{n+1}}{a_{n+1}^o} \frac{q_{N-1}^{(n+1)}(x)}{p_{N-1}^{(n+1)}(x)} \omega^{\mp N} p_n(x) .$$

For recurrence coefficients satisfying (2.2.14) we then find, using (2.2.30)

$$(2.4.10)\qquad f_{\pm}(x;n_0) = a^{0}_{n_0+1}\,\frac{q^{\pm}_{n_0}(x)}{q^{(n_0+1)}_{N-1}(x)}\,\omega^{\pm N}\psi^{\pm}_{n_0}(x)\,.$$

This leads to the following result

Theorem 2.20. Suppose that

$$\sum_{n=1}^{\infty}\left\{|b_n - b^{0}_n| + \left|1 - \left(\frac{a_n}{a^{0}_n}\right)^2\right|\right\} < \infty$$

then for every integer j

$$\lim_{k\to\infty} \omega^{-kN}\, p_{kN+j}(x) = \frac{q^{(j+1)}_{N-1}(x)f_{+}(x)}{a^{0}_{j+1}\,q^{+}_{j}(x)\,(\omega^{N} - \omega^{-N})}$$

uniformly on compact subsets of $\mathbb{C}\backslash E$.

Proof : Clearly (2.2.17) holds so that we can let n_0 tend to infinity in (2.4.10) for even $x \in \mathbb{C}\backslash E$, giving

$$(2.4.11)\qquad f_{\pm}(x) = \lim_{n_0\to\infty} \frac{a^{0}_{n_0+1}\,q^{\pm}_{n_0}(x)}{q^{(n_0+1)}_{N-1}(x)}\,\omega^{N}\psi^{\pm}_{n_0}(x)\,.$$

Let $n_0 = kN + j$, then

$$(2.4.12)\qquad f_{\pm}(x) = \frac{a^{0}_{j+1}\,q^{\pm}_{j}(x)}{q^{(j+1)}_{N-1}(x)}\lim_{k\to\infty} \omega^{(1\mp k)N}\,\psi^{\pm}_{kN+j}(x)\,.$$

Substract the expression (2.4.6) with superscript + from the expression with superscript - and multiply this by ω^{-kN}, then

$$(2.4.13)\qquad \omega^{-kN}p_{kN+j}(x) = \frac{\omega^{(1-k)N}\,\psi^{+}_{kN+j}(x) - \omega^{-(k+1)N}\,\psi^{-}_{kN+j}(x)}{\omega^{N} - \omega^{-N}}$$

This formula is the generalization of $T_n(x) = \frac{z^n + z^{-n}}{2}$ for the Chebyshev polynomials of the first kind. Notice that from (2.4.12) we find that

$$\psi^{-}_{kN+j}(x) = O(\omega^{-(k+1)N})$$

hence

$$\bar{\omega}^{(k+1)N}\,\psi^{-}_{kN+j}(x) = O(\omega^{-2(k+1)N})$$

and this goes to zero uniformly on compact subsets of $\mathbb{C}\setminus E$ since $|\omega| > 1$. The result then follows from (2.4.12).

■

On the set E we have oscillatory behaviour, given by

Theorem 2.21. Suppose that

$$\sum_{n=1}^{\infty} \left\{ |b_n - b_n^o| + \left|1 - \left(\frac{a_n}{a_n^o}\right)^2\right| \right\} < \infty$$

then

$$\lim_{k\to\infty} \left| p_{kN+j}(x)\sqrt{\frac{\pi}{2}\, a_{j+1}^{o}\,\sigma(x)\,\sqrt{4 - \left\{q_N(x) - \frac{a_N^o}{a_{N+1}^o}\, q_{N-2}^{(1)}(x)\right\}^2}\,/\,|q_{N-1}^{(j+1)}(x)|} \right.$$

$$\left. - \cos(kN\theta + \Gamma_j(\theta)) \right| = 0$$

holds uniformly for closed subsets in $E\setminus\{\omega^{2N} = 1\}$, where $\theta = \arg \omega(x)$ and $\Gamma_j(\theta) = \arg\{-if_+(x)q_j^-(x)\}$.

Proof : This follows from (2.2.24), (2.2.18), Theorem 2.4 and the properties of ω on E.

■

2.5. Weak asymptotics

In this section we present some more asymptotic results for orthogonal polynomials with asymptotically periodic recurrence coefficients. We will impose no conditions on the rate of convergence in (2.2.2). The results in this section are generalizations of results by Maki [114], Chihara [42] and Nevai [138] who studied the case $N = 1$. Chihara [43] also considered the case $N = 2$. Recall that we defined X_1 to be the accumulation points of the zeros $\{x_{j,n} : j = 1,\dots,n;\ n = 1,2,3,\dots\}$ and X_2 as the set of all zeros x_i for which $p_n(x_i) = 0$ for infinitely many n.

Theorem 2.22. Let $\{p_n(x) : n = 0,1,2,\dots\}$ be orthogonal polynomials with asymptotically period recurrence coefficients, then for the monic polynomials

$$(2.5.1)\qquad \lim_{n\to\infty} \frac{\hat{p}_{n-N}(x)}{\hat{p}_n(x)} = \frac{1}{2C^{2N}} \Big\{ \hat{q}_N(x) - (a_N^o)^2\, \hat{q}_{N-2}^{(1)}(x) - \sqrt{\{\hat{q}_N(x) - (a_N^o)^2\, \hat{q}_{N-2}^{(1)}(x)\}^2 - 4C^{2N}} \Big\}$$

uniformly on compact subsets of $\mathbb{C}\backslash(X_1 \cup X_2)$, where

$$C = \Big(\prod_{j=1}^{N} a_j^o\Big)^{1/N} .$$

Proof : Combine (2.4.1) and (2.4.4) for monic polynomials, then

$$(2.5.2)\qquad \hat{p}_{n+2N}(x)\hat{p}_{N-1}^{(n+1)}(x) = \Big\{ \hat{p}_{N-1}^{(n+1)}(x)\hat{p}_N^{(n+N)}(x) - a_{n+N+1}^2 \hat{p}_{N-1}^{(n+N+1)}(x)\hat{p}_{N-2}^{(n+1)}(x) \Big\} \hat{p}_{n+N}(x) - (a_{n+2}a_{n+3}\cdots a_{n+N+1})^2 \hat{p}_{N-1}^{(n+N+1)}(x)\hat{p}_n(x) .$$

The convergence of the recurrence coefficients to a periodic sequence implies

$$(2.5.3)\qquad \lim_{k\to\infty} \hat{p}_m^{(kN+j)}(x) = \hat{q}_m^{(j)}(x) .$$

Take $n = (k-1)N+j$ in (2.5.2) then for $\phi_k(x) = \hat{p}_{kN+j}(x)$ we find

$$A_k\phi_{k+1}(x) = B_k\phi_k(x) - C_k\phi_{k-1}(x)$$

with

$$A_k = \hat{p}_{N-1}^{((k-1)N+j+1)}(x)$$

$$B_k = \hat{p}_{N-1}^{((k-1)N+j+1)}(x)\hat{p}_N^{(kN+j)}(x) - a_{kN+j+1}^2 \hat{p}_{N-1}^{(kN+j+1)}(x)\hat{p}_{N-2}^{((k+1)N+j+1)}(x)$$

$$C_k = (a_{(k-1)N+j+2}\, a_{(k-1)N+j+3} \cdots a_{kN+j+1})^2\, \hat{p}_{N-1}^{(kN+j+1)}(x) .$$

Therefore $\{\phi_k : k = 1,2,\ldots\}$ satisfies a recurrence relation with coefficients $\{A_k,B_k,C_k : k = 1,2,\ldots\}$ that converge as k tends to infinity, namely

$$\lim_{k \to \infty} A_k = \hat{q}_{N-1}^{(j+1)}(x)$$

$$\lim_{k \to \infty} B_k = \hat{q}_{N-1}^{(j+1)}(x)\left\{\hat{q}_N^{(j)}(x) - (a_{j+1}^o)^2\hat{q}_{N-2}^{(j+1)}(x)\right\}$$

$$\lim_{k \to \infty} C_k = (a_1^o a_2^o \ldots a_N^o)^2 \hat{q}_{N-1}^{(j+1)}(x) .$$

Therefore we can use a result by Poincaré [157] to find

$$\lim_{k \to \infty} \frac{\phi_{k-1}(x)}{\phi_k(x)} = \frac{1}{2C^{2N}} \left\{\hat{q}_N^{(j)}(x) - (a_{j+1}^o)^2\hat{q}_{N-2}^{(j+1)}(x) - \sqrt{\{\hat{q}_N^{(j)}(x) - (a_{j+1}^o)^2\hat{q}_{N-2}^{(j+1)}(x)\}^2 - 4C^{2N}}\right\}$$

which holds for $x \in \mathbb{C}\setminus(X_1 \cup X_2)$. The right hand side of this equation is the same as the right hand side of (2.5.1) because of (2.1.14). The convergence is uniform on compact subsets of $\mathbb{C}\setminus(X_1 \cup X_2)$ since

$$\frac{\phi_{k-1}(x)}{\phi_k(x)} = \prod_{i=1}^{N} \frac{\hat{p}_{(k-1)N+j+i-1}(x)}{\hat{p}_{(k-1)N+j+i}(x)}$$

and for every function in the product we have by (0.2.10)

$$\left|\frac{\hat{p}_{n-1}(x)}{\hat{p}_n(x)}\right| < \sum_{j=1}^{n} \frac{a_{j,n}}{|x - x_{j,n}|}$$

If K is a compact set in $\mathbb{C}\setminus(X_1 \cup X_2)$ then K will contain at most a finite number of zeros, therefore there exists an n_0 such that $\{p_n(x) : n > n_0\}$ has no zeros in K. Choose

$$\delta = \inf\{|z - x| : z \in K,\ x \in X_1 \cup X_2 \cup Z_{n_0}\}$$

with $Z_{n_0} = \{x_{j,n} : j = 1,\ldots,n;\ n > n_0\}$ then $\delta > 0$ and

$$\left|\frac{\phi_{k-1}(x)}{\phi_k(x)}\right| < \frac{1}{\delta^N}$$

for k large enough and $x \in K$. Consequently ϕ_{k-1}/ϕ_k is uniformly bounded on K for k large enough so that the uniform convergence follows by Stieltjes-Vitali's theorem. ∎

This result has some immediate consequences :

Theorem 2.23. For orthogonal polynomials with asymptotically periodic recurrence coefficients we have

(2.5.4) $$\lim_{n\to\infty} \frac{1}{n}\frac{\hat{p}_n'(x)}{\hat{p}_n(x)} = \frac{1}{N}\frac{\left\{\hat{q}_N(x) - (a_N^o)^2\hat{q}_{N-2}^{(1)}(x)\right\}'}{\sqrt{\{\hat{q}_N(x) - (a_N^o)^2\hat{q}_{N-2}^{(1)}(x)\}^2 - 4C^{2N}}},$$

and

(2.5.5) $$\lim_{n\to\infty} |\hat{p}_n(x)|^{1/n} = \left|\frac{1}{2}\left\{\hat{q}_N(x) - (a_N^o)^2\hat{q}_{N-2}^{(1)}(x) + \sqrt{\{\hat{q}_N(x) - (a_N^o)^2\hat{q}_{N-2}^{(1)}(x)\}^2 - 4C^{2N}}\right\}\right|^{1/N},$$

uniformly on compact subsets of $\mathbb{C}\backslash(X_1 \cup X_2)$.

Proof : We can take derivatives on both sides of (2.5.1) because the convergence is uniformly on compact subsets of $\mathbb{C}\backslash(X_1 \cup X_2)$. Denote the right hand side of (2.5.1) by Q(x), then this gives

$$\lim_{n\to\infty} \frac{\hat{p}_{n-N}(x)}{\hat{p}_n(x)}\left\{\frac{\hat{p}_{n-N}'(x)}{\hat{p}_{n-N}(x)} - \frac{\hat{p}_n'(x)}{\hat{p}_n(x)}\right\} = Q'(x)$$

uniformly on compact subsets of $\mathbb{C}\backslash(X_1 \cup X_2)$. If we use (2.5.1) again, then

$$\lim_{n\to\infty} \left\{\frac{\hat{p}_{n-N}'(x)}{\hat{p}_{n-N}(x)} - \frac{\hat{p}_n'(x)}{\hat{p}_n(x)}\right\} = \frac{Q'(x)}{Q(x)}.$$

Use Cesàro's lemma, then

$$\frac{1}{n}\frac{\hat{p}_n'(x)}{\hat{p}_n(x)} = \frac{1}{n}\sum_{i=o}^{[\frac{n}{N}]}\left\{\frac{\hat{p}_{n-iN}'(x)}{\hat{p}_{n-iN}(x)} - \frac{\hat{p}_{n-(i+1)N}'(x)}{\hat{p}_{n-(i+1)}(x)}\right\} \longrightarrow -\frac{1}{N}\frac{Q'(x)}{Q(x)}$$

uniformly on compact subsets of $\mathbb{C}\backslash(X_1 \cup X_2)$. If we calculate Q'(x)/Q(x) explicitely, then (2.5.4) follows. In order to prove (2.5.5) we notice that

$$|\hat{p}_n(x)| = \prod_{k=0}^{[\frac{n}{N}]} \left|\frac{\hat{p}_{n-kN}(x)}{\hat{p}_{n-(k+1)N}(x)}\right|$$

so that Cesàro's lemma implies (2.5.5). ■

These theorems have some interesting applications. Let ν_n be the discrete measure given by (1.2.1), then the Stieltjes transform of ν_n is given by

$$S(\nu_n;x) = \int \frac{1}{x-t}\, d\nu_n(t) = \frac{1}{n}\frac{\hat{p}_n'(x)}{\hat{p}_n(x)} .$$

Then (2.5.4) implies

$$\lim_{n\to\infty} S(\nu_n;x) = \frac{1}{N}\frac{\left\{\hat{q}_N(x) - (a_N^o)^2\hat{q}_{N-2}^{(1)}(x)\right\}'}{\sqrt{\{q_N(x) - (a_N^o)^2 q_{N-2}^{(1)}(x)\}^2 - 4C^{2N}}}$$

uniformly on compact subsets of $\mathbb{C}\setminus(X_1 \cup X_2)$. By the Grommer-Hamburger theorem (Appendix) this limit is the Stieltjes transform of some probability measure. If we invert this Stieltjes transform by using the inversion formula (A.1.2) then we find the measure

$$d\mu_E(x) = \frac{1}{N\pi}\frac{\left|\left\{\hat{q}_N(x) - (a_N^o)^2\, q_{N-2}^{(1)}(x)\right\}'\right|}{\sqrt{4C^{2N} - \{\hat{q}_N(x) - (a_N^o)^2\hat{q}_{N-2}^{(1)}(x)\}^2}}\,dx \qquad x \in E\setminus\{\omega^{2N} = 1\}$$

which is exactly the equilibrium measure on the set E. As a result we have for every continuous function f

$$\lim_{n\to\infty} \frac{1}{n}\sum_{j=1}^{n} f(x_{j,n}) = \int_E f(x)\,d\mu_E(x)$$

and this can be generalized to bounded measurable functions f with discontinuities on a set of μ_E-measure zero, which means that the result holds for Riemann integrable functions.

<u>Theorem 2.24.</u> If the recurrence coefficients of orthogonal polynomials are asymptotically periodic, then

$$\text{(2.5.6)} \qquad \lim_{k\to\infty} \frac{\hat p_{kN+j-1}(x)}{\hat p_{kN+j}(x)} = \frac{1}{2(a_j^o)^2\hat q_{N-1}^{(j+1)}(x)} \Big\{\hat q_N^{(j)}(x) + (a_j^o)^2 q_{N-2}^{(j+1)}(x) - \sqrt{\{\hat q_N(x) - (a_N^o)^2\hat q_{N-2}^{(1)}(x)\}^2 - 4C^{2N}}\Big\}$$

uniformly on compact subsets of $\mathbb{C}\setminus(X_1 \cup X_2)$.

Proof : Consider (2.4.2) for monic polynomials with $n = kN + j$ and $n = (k+1)N+j$, then

$$\hat p_{(k+1)N+j}(x) = \hat p_{kN+j}(x)\hat p_N^{(kN+j)}(x) - a_{kN+j}^2\hat p_{kN+j-1}(x)\hat p_{N-1}^{(kN+j+1)}(x) .$$

Divide this by $\hat p_{kN+j}(x)$ then by Theorem 2.22, (2.5.3) and (2.2.2) we find

$$\lim_{k\to\infty} \frac{\hat p_{kN+j-1}(x)}{\hat p_{kN+j}(x)} = \frac{\hat q_N^{(j)}(x) - \frac{1}{Q(x)}}{(a_j^o)^2\hat q_{N-1}^{(j+1)}(x)}$$

where Q is the right hand side of (2.5.1), proving the result.

■

The right hand side of (2.5.6) depends on j and therefore the ratio $\hat p_{n-1}(x)/\hat p_n(x)$ has N different limits. The limits in (2.5.1), (2.5.4), (2.5.5) and (2.5.6) only depend on the limit sequences $\{a_{n+1}^o, b_n^o : n = 0,1,2,...\}$ and not on the rate at which (2.2.2) holds. Theorems 2.22 - 2.24 are therefore invariance theorems giving the same limits for a large class of orthogonal polynomials.

Define a new sequence of measures $\{\lambda_n : n = 1,2,...\}$ by

$$\begin{cases} \lambda_n(\{x_{j,n}\}) = \lambda_{j,n}p_{n-1}^2(x_{j,n}) & j = 1,2,...,n \\ \lambda_n(A) = 0 & \text{A contains no zeros of } p_n \end{cases}$$

then the Stieltjes transform of λ_n is, by (0.2.10) and (0.2.11)

$$S(\lambda_{kN+j};x) = \int \frac{1}{x-t}\, d\lambda_{kN+j}(t) = \frac{\hat p_{kN+j-1}(x)}{\hat p_{kN+j}(x)}$$

and by (2.5.6) and the theorem of Grommer-Hamburger (appendix) it follows that

$\{\lambda_{kN+j} : k = 1,2,3,\ldots\}$ converges weakly to a probability measure $\lambda^{(j)}$ with Stieltjes transform given by the right hand side of (2.5.6). This Stieltjes transform may have poles at the zeros of $q_{N-1}^{(j+1)}$ and therefore the measure $\lambda^{(j)}$ may have positive mass at the zeros of $q_{N-1}^{(j+1)}$. By Lemma 2.2 these zeros are all in the set F (the open intervals between the intervals of E). As a consequence we have

$$\lim_{k\to\infty} \sum_{i=1}^{kN+j} \lambda_{i,kN+j} p_{kN+j-1}^2(x_{i,kN+j}) f(x_{i,kN+j}) = \int f(x)\,d\lambda^{(j)}(x)$$

which holds for every bounded measurable function f with discontinuities on a set of $\lambda^{(j)}$-measure zero. In particular f should have no discontinuities at those zeros of $q_{N-1}^{(j)}$ where $\lambda^{(j)}$ has positive mass. The measures $\lambda^{(j)}$ are calculated explicitely for N = 1 and N = 2 in [189].

The results in this section remain valid when one of the accumulation points of the sequence $\{a_n : n = 1,2,\ldots\}$ is equal to zero. In that case we have

$$C = \left(\prod_{j=1}^{N} a_j^o\right)^{1/N} = 0 .$$

If $a_m^o = 0$ then Theorems 2.22 - 2.24 become respectively

$$\lim_{n\to\infty} \frac{\hat{p}_{n-N}(x)}{\hat{p}_n(x)} = \frac{1}{\hat{q}_N(x) - (a_N^o)^2 \hat{q}_{N-2}^{(1)}(x)} = \frac{1}{\hat{q}_N^{(m)}(x)}$$

$$\lim_{n\to\infty} \frac{1}{n} \frac{\hat{p}_n'(x)}{\hat{p}_n(x)} = \frac{1}{N} \frac{\{\hat{q}_N(x) - (a_N^o)^2 \hat{q}_{N-2}^{(1)}(x)\}'}{\hat{q}_N(x) - (a_N^o)^2 \hat{q}_{N-2}^{(1)}(x)} = \frac{1}{N} \frac{(\hat{q}_N^{(m)}(x))'}{\hat{q}_N^{(m)}(x)}$$

$$\lim_{n\to\infty} |\hat{p}_n(x)|^{1/n} = |\hat{q}_N(x) - (a_N^o)^2 \hat{q}_{N-2}^{(1)}(x)|^{1/N} = |\hat{q}_N^{(m)}(x)|^{1/N}$$

$$\lim_{k\to\infty} \frac{\hat{p}_{kN+j-1}(x)}{\hat{p}_{kN+j}(x)} = \frac{\hat{q}_{N-2}^{(j+1)}(x)}{\hat{q}_{N-1}^{(j+1)}(x)}$$

uniformly on compact subsets of $\mathbb{C}\backslash(X_1 \cup X_2)$. These limits, except for the third, are rational functions. This means that the weak limits of the measures $\{\nu_n : n = 1,2,\ldots\}$ and $\{\lambda_n : n = 1,2,\ldots\}$ are discrete measures with a finite support. This property was already mentioned by Krein [103]. Some examples are given in [189].

2.6. Periodicity of period N=1

Let us consider the special case N = 1 in more detail. The condition (2.2.2) now becomes

$$\lim_{n\to\infty} a_n = a > 0 \quad , \quad \lim_{n\to\infty} b_n = b \in \mathbb{R}$$

and the set E in (2.1.21) is the interval [b-2a,b+2a]. The polynomials $\{p_n(\frac{x-b}{2a}) : n = 0,1,2,\ldots\}$ are again orthogonal polynomials with recurrence coefficients that now satisfy

$$(2.6.1) \qquad \lim_{n\to\infty} a_n = \frac{1}{2} \quad , \quad \lim_{n\to\infty} b_n = 0$$

so that we may consider a = 1/2 and b = 0 without loss of generality. The essential spectrum then is the interval [-1,1]. Orthogonal polynomials with recurrence coefficients as in (2.6.1) have been considered in great detail by Nevai [138] and many others. Theorem 2.14 for this case becomes

Theorem 2.25 (Geronimus [79], Dombrowski-Nevai [50]) Let $m \geq 0$ be a fixed integer and let $\{b_k \in \mathbb{R} : k = 0,1,\ldots\}$ and $\{a_k > 0 : k = 1,2,\ldots\}$ be recurrence coefficients such that $b_k = 0$ $(k \geq m)$ and $a_k = 1/2$ $(k > m)$. Let $\{p_n(x) : n = 0,1,2,\ldots\}$ be the orthogonal polynomials generated by these recurrence coefficients, then the spectral measure μ is given by

$$(2.6.2) \qquad d\mu(x) = \sigma(x)dx + \sum_i \rho_i dU(x - x_i)$$

where $\sigma(x)$ is a weight function on [-1,1] given by

$$(2.6.3) \qquad \sigma(x) = \frac{2}{\pi} \frac{\sqrt{1-x^2}}{p_m^2(x) - 2xp_m(x)p_{m+1}(x) + p_{m+1}^2(x)} ,$$

the x_i are those zeros of $S_m(x) = p_m^2(x) - 2xp_m(x)p_{m+1}(x) + p_{m+1}^2(x)$ for which $|p_{m+1}(x_i)| < |p_m(x_i)|$ and the corresponding mass ρ_i is given by

$$(2.6.4) \qquad \rho_i = -\frac{\sqrt{x_i^2 - 1}}{S_m'(x_i)}$$

The general case (2.6.1) has been dealt with by Geronimo and Case [72] and Nevai [138]. The asymptotic behaviour of the orthogonal polynomials can be found by combining their results :

Theorem 2.26. Suppose that

$$(2.6.5) \qquad \sum_{n=2}^{\infty} \log n \{|1 - 4a_n^2| + |b_n|\} < \infty$$

then the spectral measure μ is given by

$$d\mu(x) = \sigma(x)dx + \sum_i \rho_i dU(x - x_i)$$

where $\sigma(x)$ is a weight function on [-1,1] and x_i are mass points outside (-1,1). The asymptotic result

$$(2.6.6) \qquad \lim_{n \to \infty} \frac{p_n(x)}{(x + \sqrt{x^2 - 1})^n} = \frac{R(x)}{\sqrt{2\pi}} \exp\left\{ - \frac{\sqrt{x^2 - 1}}{2\pi} \int_{-1}^{1} \frac{\log(\sigma(t)\sqrt{1 - t^2})}{\sqrt{1 - t^2}} \frac{dt}{x - t} \right\}$$

holds uniformly on closed subsets of $\bar{\mathbb{C}} \setminus [-1,1]$, with

$$R(x) = \frac{\prod_{z_i > 0} (z_i - z) \prod_{z_j < 0} (z - z_j)}{\prod_i (1 - z z_i)}$$

$$z = x - \sqrt{x^2 - 1} \quad , \qquad z_i = x_i - \sqrt{x_i^2 - 1} \, .$$

If

$$\sum_{n=1}^{\infty} n\{|1 - 4a_n^2| + |b_n|\} < \infty$$

then the number of mass points x_i is finite.

This result is very similar to the asymptotic behaviour of orthogonal polynomials in the Szegö class on [-1,1] (Theorem 1.7). The reason for this is that, by Theorem 2.17, the weight function σ belongs to the Szegö class on [-1,1] when (2.6.5) is valid. Notice that the right hand side of (2.6.6) is actually a factorization into a Blaschke product R (the discrete part of μ) and an inner function dealing with the absolutely continuous part of μ. Condition (2.6.5) is not the most general condition in order that σ belongs to the Szegö class on [-1,1]. Nevai [138],[141] showed that when $\text{supp}(\mu) = [-1,1]$ it is sufficient that the coefficients satisfy

$$(2.6.7) \qquad \sum_{n=1}^{\infty} \{|1 - 4a_n^2| + |b_n|\}$$

in order that σ belongs to the Szegö class.
Máté and Nevai have analyzed the weight function σ more carefully under weaker conditions on the recurrence coefficients.

<u>Theorem 2.27.</u> (Máté-Nevai [117]) Suppose that (2.6.1) holds together with

$$(2.6.8) \qquad \sum_{n=1}^{\infty} \{|a_{n+1} - a_n| + |b_{n+1} - b_n|\} < \infty$$

then the distribution function of the spectral measure μ is continuously differentiable in (-1,1) and $\sigma(x) = d\mu/dx$ is strictly positive for $-1 < x < 1$.

Máté-Nevai-Totik subsequently used the same condition (2.6.8) and the same techniques to find the asymptotic behaviour of the orthogonal polynomials.

<u>Theorem 2.28.</u> (Máté-Nevai-Totik [123], Van Assche-Geronimo [194]). Suppose that (2.6.1) and (2.6.8) hold and let

$$t_k(x) = \frac{x - b_k}{2a_{k+1}} + \sqrt{\left(\frac{x - b_k}{2a_{k+1}}\right)^2 - \frac{a_k}{a_{k+1}}}$$

then there exists a function g such that

$$(2.6.9) \qquad \lim_{n \to \infty} \frac{p_n(x)}{\prod_{k=1}^{n} t_k(x)} = \frac{g(x)}{2\sqrt{x^2 - 1}}$$

uniformly on compact sets of $\mathbb{C}\setminus[-1,1]$ and

$$(2.6.10) \qquad \lim_{n \to \infty} \left\{\sqrt{\sigma(x)}\sqrt{1 - x^2}\, p_n(x) - \sqrt{\tfrac{2}{\pi}} \sin\left(\sum_{k=1}^{n} \arg t_k(x) + \arg g(x)\right)\right\} = 0$$

uniformly on compact subsets of (-1,1). The function g is continuous in (-1,1) and if K is a compact set in $\mathbb{C}\setminus[-1,1]$ then $g(x) \prod_{k=1}^{N} t_k(x)$ is analytic in K for large enough N. Moreover

$$g(x) = \lim_{y \to 0+} g(x + iy) \qquad x \in (-1,1) ,$$

$g(x) = 0$ if and only if $x \in \mathrm{supp}(\mu)\setminus[-1,1]$ and $g'(x)$ exists and is different from zero if $g(x) = 0$. If $K \subset \mathrm{supp}(\mu)\setminus[-1,1]$ is a compact set then

$$\lim_{n \to \infty} p_n(x) \prod_{k=1}^{n} t_k(x) = \frac{2}{\mu(\{x\})\, g'(x)} \tag{2.6.11}$$

uniformly in K.

<u>Sketch of the proof</u> : Define $s_k(x)$ by

$$s_k(x) = \frac{a_k}{a_{k+1}} \frac{1}{t_k(x)}$$

and introduce

$$\phi_n(x) = p_n(x) - s_n(x) p_{n-1}(x) \tag{2.6.12}$$

then one easily finds

$$\phi_{n+1}(x) - t_n(x)\phi_n(x) = (s_n(x) - s_{n+1}(x)) p_n(x) . \tag{2.6.13}$$

If we set $\varepsilon_n = |a_{n+1} - a_n| + |b_{n+1} - b_n|$ then some elementary calculus gives

$$|t_n - t_{n+1}| = O(\varepsilon_n + \varepsilon_{n+1}) \quad , \quad |s_n - s_{n+1}| = O(\varepsilon_n + \varepsilon_{n+1}) \tag{2.6.14}$$

uniformly on compact subsets of $\mathbb{C}\setminus[-1,1]$. From (2.6.13) we obtain

$$\phi_{n+1}(x) = t_n(x)\phi_n(x) \left\{ 1 + \frac{s_n - s_{n+1}}{t_n(x) - \frac{a_n}{a_{n+1}} \frac{p_{n-1}(x)}{p_n(x)}} \right\}. \tag{2.6.15}$$

Thus, by (2.6.14), (2.6.8) and by

$$\lim_{n \to \infty} \frac{p_{n-1}(x)}{p_n(x)} = \frac{1}{x + \sqrt{x^2 - 1}} \quad , \quad \lim_{n \to \infty} t_n(x) = x + \sqrt{x^2 - 1}$$

(see Theorem 2.22) we find that there exist a function G such that

$$\lim_{n \to \infty} \prod_{k=N}^{n} \frac{\phi_{k+1}(x)}{t_k(x)\phi_k(x)} = G \neq 0 \tag{2.6.16}$$

uniformly on a compact set K in $\mathbb{C}\setminus[-1,1]$, with N large enough. Setting $g(x) = \phi_N(x)G(x)/\prod_{k=1}^{N-1} t_k(x)$ gives

$$(2.6.17) \qquad \lim_{n\to\infty} \frac{p_n(x) - s_n(x)p_{n-1}(x)}{\prod_{k=1}^{n-1} t_k(x)} = g(x)$$

uniformly on compact sets in $\mathbb{C}\setminus[-1,1]$. If K is a compact set in (-1,1) then

$$\left|t_n(x) - \frac{a_n}{a_{n+1}} \frac{p_{n-1}(x)}{p_n(x)}\right| \geq \left|\mathrm{Im}\left\{t_n(x) - \frac{a_n}{a_{n+1}} \frac{p_{n-1}(x)}{p_n(x)}\right\}\right|$$

$$= |\mathrm{Im}\ t_n(x)| > \delta$$

with δ strictly positive so that (2.6.16) also holds on compact sets $K \subset (-1,1)$, giving

$$(2.6.18) \qquad p_n(x) - s_n(x)p_{n-1}(x) = \prod_{k=1}^{n-1} t_k(x)\ g(x) + o(1)$$

The result in (2.6.9) follows from (2.6.17) and (2.6.10) follows from (2.6.18) by taking the imaginary part of both sides of the equation (2.6.18) and by

$$|g(x)| \prod_{k=1}^{\infty} |t_k(x)| = \sqrt{\frac{2}{\pi} \frac{\sqrt{1-x^2}}{\sigma(x)}}$$

(Máté-Nevai [117]). The continuity of the function g in $\{z \in \mathbb{C} : \mathrm{Im}\ z \geq 0, -1 < \mathrm{Re} z < 1\}$ has been proved in Van Assche-Geronimo [194]. The asymptotic behaviour in (2.6.11) follows in a similar way but instead of using (2.6.12) we introduce

$$\phi_n^*(x) = p_n(x) - t_n(x)p_{n-1}(x) .$$

Details can be found in Máté-Nevai-Totik [123].

∎

If one doesn't assume a rate on the asymptotic behaviour in (2.6.1) then the only thing that one can say about the spectral measure μ is that $\mu = \mu_c + \mu_d$, where μ_c is continuous in (-1,1) and μ_d is discrete with at most a denumerable amount of mass points with accumulation points in [-1,1]. This result is due to Blumenthal

[29] and corresponds to Theorem 2.26 for this special case. Without any extra assumptions on the recurrence coefficients a_n and b_n one can not conclude that μ has an absolutely continuous part in (-1,1). A counterexample is the following (Delyon-Simon-Souillard [49]). Let $a_n = 1/2$ $(n = 1,2,\ldots)$ and $b_n = c_n X_n$, where

$$C_1 n^{-\alpha} < c_n < C_2 n^{-\alpha} \qquad n = 1,2,\ldots$$

(C_1, C_2 are two positive constants) and $\{X_0, X_1, X_2, \ldots\}$ are independent and identically distributed random variables. If $\alpha > 0$ then (2.6.1) is satisfied but under fairly general conditions on the probability distribution of X one can show that for $0 < \alpha < 1/2$ the spectral measure is purely discrete and the mass points are dense in [-1,1].

An inverse problem consists of finding conditions on the spectral measure μ under which the recurrence coefficients a_n and b_n satisfy (2.6.1). An important result, but probably not the best possible, is the following.

Theorem 2.29. (Rakhmanov [161],[162]; Máté-Nevai [116]).

Suppose that $\text{supp}(\mu) = [-1,1]$ and that $d\mu/dx$ is positive almost everywhere in [-1,1] then (2.6.1) holds.

CHAPTER 3 : PROBABILISTIC PROOFS OF ASYMPTOTIC FORMULAS

Wolfgang Stadje gave in [172] a probabilistic proof for asymptotic formulas involving Bessel functions : some Bessel functions can be written as the probability distribution of a partial sum of symmetrized independent and identically distributed random variables. Therefore one can use limit theorems in probability theory to obtain asymptotic formulas for these Bessel functions. In this chapter the same idea will be used to obtain asymptotic formulas for some classical and semi-classical polynomials.

3.1. The local limit theorem in probability theory

In this section we introduce some notions of probability theory needed for this chapter. Let Z be a *random variable*, defined on a probability space $(\Omega,\mathcal{B},P)$. The distribution function of this random variable is

$$F_Z(t) = P(Z < t) .$$

The quantities E(Z) and Var(Z) are respectively the *mean* (*expectation*) and the *variance* of the random variable Z and are defined by

$$E(Z) = \int_{-\infty}^{\infty} x dF_Z(x) \quad , \qquad Var(Z) = \int_{-\infty}^{\infty} (x - E(Z))^2 dF_Z(x) .$$

These two notions only makes sense when the integrals involved are convergent. If the random variable Z only takes integer values then the *span* H of Z (or the distribution F_Z) is defined as

$$H = \max\{k \in \mathbb{Z}^+ : supp(F_Z) \subset k\mathbb{Z} + m \text{ for some } m\} .$$

The next theorem, which is classical in probability theory, is the *local limit theorem* for lattice distributions :

Theorem 3.1. (Petrov [154], p. 189) : Let $\{Z_i : i = 1,2,\ldots\}$ be a sequence of independent random variables, defined on a probability space $(\Omega,\mathcal{B},P)$ and with values in $\mathbb{Z}$. If there are k different distribution functions $\{V_1,V_2,\ldots,V_k\}$ in the sequence $\{F_i : i = 1,2,\ldots\}$ of distribution functions corresponding to

$\{Z_i : i = 1,2,\ldots\}$ then we denote by $\{W_1,\ldots,W_h\}$ those distribution functions in $\{V_1,\ldots,V_k\}$ that are not degenerate (H is finite) and that appear infinitely often in the sequence $\{F_i : i = 1,2,\ldots\}$ ($h \leq k$ and we suppose $h \geq 1$). If H_r is the span of W_r $(r = 1,\ldots,h)$ then

$$\text{(3.1.1)} \qquad \sup_N \left| s_n P(Z_1 + Z_2 + \ldots + Z_n = N) - \frac{1}{\sqrt{2\pi}} \exp\left\{ - \frac{(N - m_n)^2}{2s_n^2} \right\} \right| \to 0 \quad (n \to \infty)$$

if and only if

$$\text{(3.1.2)} \qquad \text{g.c.d.}\ (H_1, H_2, \ldots, H_h) = 1$$

where

$$m_n = E(Z_1 + Z_2 + \ldots + Z_n) \quad , \quad s_n^2 = \mathrm{Var}(Z_1 + Z_2 + \ldots + Z_n) \ .$$

This theorem can be improved in the sense that there exists an asymptotic expansion for (3.1.1). We need some more notation. The *characteristic function* (Fourier-Stieltjes transform) of a random variable Z is given by

$$\varphi_Z(t) = E(e^{itZ}) = \int_{-\infty}^{\infty} e^{itx} dF_Z(x) \ .$$

If we take the logarithm of φ_Z and expand this around $t = 0$,

$$\log \varphi_Z(t) = \sum_{j=1}^{\infty} \gamma_j(Z) \frac{(it)^j}{j!}$$

then $\{\gamma_j(Z) : j = 1,2,\ldots\}$ are the *cumulants* of Z (provided they exist). They have the property that

$$\gamma_j(Z_1 - Z_2) = \gamma_j(Z_1) + (-1)^j \gamma_j(Z_2)$$

whenever Z_1 and Z_2 are independent. The asymptotic expansion for the local limit theorem is then given by

<u>Theorem 3.2.</u> (Petrov [154], p. 205). Let $\{Z_i : i = 1,2,\ldots\}$ be a sequence of independent and identically distributed random variables, defined on a probability space $(\Omega,\mathcal{B},P)$ and taking values in $\mathbb{Z}$. Suppose that $\mathrm{Var}(Z_1) = \sigma^2 > 0$ and $E|Z_1|^k < \infty$ for some $k \geq 3$ and that the span of Z_1 is equal to one, then

$$(3.1.3)\qquad \sigma\sqrt{n}\,P(Z_1+Z_2+\ldots+Z_n=N)=\frac{1}{\sqrt{2\pi}}e^{-y^2/2}\left\{1+\sum_{\nu=1}^{k-2}\frac{g_\nu(y)}{n^{\nu/2}}\right\}+o(n^{-\frac{k-2}{2}}),$$

with

$$y=\frac{N-nE(Z_1)}{\sigma\sqrt{n}}$$

$$g_\nu(y)=\sum{}^{*}H_{\nu+2s}(y)\prod_{j=1}^{\nu}\frac{1}{h_j!}\left\{\frac{\gamma_{j+2}(Z_1)}{(j+2)!\sigma^{j+2}}\right\}^{h_j},$$

$\gamma_j(Z_1)$ is the jth cumulant of Z_1, the sum $\sum^{*}$ is taken over all non-negative integer solutions $(h_1,h_2,\ldots,h_\nu)$ of the equations $h_1+2h_2+\ldots+\nu h_\nu=\nu$ and $h_1+h_2+\ldots+h_\nu=s$, and H_m is the modified Hermite polynomial of degree m given by

$$(3.1.4)\qquad H_m(z)=(-1)^m e^{z^2/2}\frac{d^m}{dz^m}(e^{-z^2/2})=2^{-m/2}H_m^{(0)}\left(\frac{z}{\sqrt{2}}\right).$$

A probability distribution that we will encounter is the *Bernoulli distribution* with parameter p $(0<p<1)$, given by

$$P(Z=0)=1-p\quad,\qquad P(Z=1)=p$$

with cumulants

$$(3.1.5)\qquad \gamma_j(Z)=K_j(p)=p\sum_{k=1}^{j}(-p)^{k-1}(k-1)!\,S(j,k)$$

where $\{S(j,k) : k=1,\ldots,j;\ j=1,2,\ldots\}$ are the *Stirling numbers of the second kind*. If $\{Z_i : i=1,2,\ldots\}$ are independent random variables all having a Bernoulli distribution with the same parameter p, then $Z_1+Z_2+\ldots+Z_n$ has a *binomial distribution* with parameters n and p :

$$P(Z_1+Z_2+\ldots+Z_n=k)=\binom{n}{k}p^k(1-p)^{n-k}\qquad k=0,1,2,\ldots,n.$$

Another important distribution is de *Poisson distribution* with parameter λ $(\lambda>0)$, given by

$$P(Z=k)=e^{-\lambda}\frac{\lambda^k}{k!}\qquad k=0,1,2,\ldots.$$

with cumulants

$$\gamma_j(Z) = \lambda.$$

If $\{Z_i : i = 1,2,\ldots\}$ are independent, all having a Poisson distribution with the same parameter λ, then $Z_1 + Z_2 + \ldots + Z_n$ has a Poisson distribution with parameter $n\lambda$. Finally we mention the *negative-binomial distribution* (or *Pascal distribution*) with parameters α and p ($\alpha > 0$, $0 < p < 1$), given by

$$P(Z = k) = (-1)^k \binom{-\alpha}{K} p^\alpha (1 - p)^k \qquad k = 0,1,2,\ldots$$

and with cumulants

$$\gamma_1(Z) = K_1^*(\alpha,p) = \alpha \frac{1-p}{p} ,$$

$$\gamma_j(Z) = K_j^*(\alpha,p) = \alpha \sum_{k=1}^{j} p^{-k}(k-1)!(-1)^{k+j} S(j,k) \qquad j = 2,3,\ldots .$$

If $\{Z_i : i = 1,2,\ldots\}$ are independent random variables that all have a negative binomial distribution with parameters α and p, then $Z_1 + Z_2 + \ldots + Z_n$ has a negative-binomial distribution with parameters $n\alpha$ and p.

3.2. First order results

In this section we will frequently use the notation $a_n \sim b_n$ which means that the ratio a_n/b_n converges to one as n tends to infinity.

We will begin with a proof of an asymptotic formula for Jacobi polynomials $\{P_n^{(\alpha,\beta)}(x) : n = 0,1,2,\ldots\}$, defined in (0.1.6). Asymptotic formulas for these polynomials have been known for a long time an can be found in Szegö's book ([175], Theorem 8.21.7). The standard proofs use the differential equation combined with a steepest descend method. We will give a short proof for an asymptotic formula in case the parameters α and β are integers and $x > 1$. The result in the following Theorem corresponds to the asymptotic formulas (1.3.5) and (1.3.6).

Theorem 3.3. (Maejima-Van Assche [111]). Suppose that α and β are non-negative integers and $x > 1$, then for Jacobi polynomials $\{P_n^{(\alpha,\beta)}(x) : n = 0,1,2,\ldots\}$ we have

$$P_n^{(\alpha,\beta)}(x) \sim \frac{1}{\sqrt{2\pi n}} (x-1)^{-\alpha/2}(x+1)^{-\beta/2}(\sqrt{x+1} + \sqrt{x-1})^{\alpha+\beta}$$

$$(x^2 - 1)^{-1/4} (x + \sqrt{x^2 - 1})^{n+1/2} .$$

Proof : Suppose that $\beta > \alpha$ (the other case can be handled similarly). Let

$$x = \frac{t^2+1}{t^2-1} \qquad (t > 1)$$

then by (0.1.6) we have

$$(3.2.1) \qquad P_n^{(\alpha,\beta)}(x) = \frac{1}{(t^2-1)^n} \sum_{j=0}^{n} \binom{n+\alpha}{j}\binom{n+\beta}{n-j} t^{2j} .$$

Consider two independent sequences $\{X_i : i = 1,2,\ldots\}$ and $\{Y_i : i = 1,2,\ldots\}$ of independent Bernoulli random variables with parameter p and put $S_n = X_1 + \ldots + X_n$, $S_n^* = Y_1 + \ldots + Y_n$, then

$$P(S_{n+\beta} - S_{n+\alpha}^* = \beta) = \sum_{j=0}^{n} P(S_{n+\beta} = \beta+j)P(S_{n+\alpha}^* = j)$$

$$= p^\beta(1-p)^{\alpha+2n} \sum_{j=0}^{n} \binom{n+\alpha}{j}\binom{n+\beta}{n-j}\left(\frac{p}{1-p}\right)^{2j} .$$

If we let p be equal to $\frac{t}{1+t}$ $(\frac{1}{2} < p < 1)$ then by (3.2.1)

$$(3.2.2) \qquad P_n^{(\alpha,\beta)}(x) = \frac{p^{-\beta}(1-p)^{-2n-\alpha}}{(t^2-1)^n} P(S_{n+\beta} - S_{n+\alpha}^* = \beta) .$$

Clearly $S_{n+\beta} - S_{n+\alpha}^*$ has the same distribution as $\sum_{j=1}^{n+\beta} Z_j$, with

$$Z_j = \begin{cases} X_j & 1 \leq j \leq \beta - \alpha \\ X_j - Y_j & \beta - \alpha + 1 \leq j \leq n + \beta . \end{cases}$$

The sequence $\{Z_j : j = 1,2,\ldots\}$ satisfies condition (3.1.2) and a simple calculation yields

$$m_{n+\beta} = E(S_{n+\beta} - S_{n+\alpha}^*) = (\beta - \alpha)p$$

$$s_{n+\beta}^2 = Var(S_{n+\beta} - S_{n+\alpha}^*) = (2n + \alpha + \beta)p(1 - p) .$$

The local limit theorem can be used and from (3.1.1) we find

$$\lim_{n\to\infty} \left| \sqrt{(2n+\alpha+\beta)p(1-p)}\, P(S_{n+\beta} - S^*_{n+\alpha} = \beta) - \frac{1}{\sqrt{2\pi}} \exp\left\{ - \frac{(\beta-(\beta-\alpha)p)^2}{2(2n+\alpha+\beta)p(1-p)} \right\} \right| = 0$$

from which we find

$$P(S_{n+\beta} - S^*_{n+\alpha} = \beta) \sim \frac{1}{2\sqrt{n\pi p(1-p)}} .$$

Use this in (3.2.2) together with the formulas

$$p = \frac{\sqrt{x+1}}{\sqrt{x+1} + \sqrt{x-1}} \quad ; \quad 1 - p = \frac{\sqrt{x-1}}{\sqrt{x+1} + \sqrt{x-1}}$$

$$t^2 - 1 = \frac{2}{x-1} \quad ; \quad (\sqrt{x+1} + \sqrt{x-1})^2 = 2(x + \sqrt{x^2-1})$$

then the result follows.

∎

The second result is about (generalized) Laguerre polynomials $\{L_n^{(\alpha)}(x) : n = 0,1,2,\ldots\}$ defined in (0.1.9). Asymptotic results for these polynomials have been given by Moecklin [135] and Erdélyi [51]. We will give a short proof of the Plancherel-Rotach formula when α is an integer and $x < 0$. The case $x > 4n$ has been treated by Taylor [178] and can also been found in Szegö's book ([175], Theorem 8.22.8b).

<u>Theorem 3.4</u> (Maejima-Van Assche [111]). Let α be a non-negative integer and $x < 0$ then

(i) $$L_n^{(\alpha)}(nx) \sim \frac{1}{\sqrt{2\pi n}} \left(\frac{-x + \sqrt{x^2 - 4x}}{2}\right)^{2n+\alpha+1} (-x)^{-n-\alpha-1/2} \times \exp\left\{\frac{n}{2}(x + \sqrt{x^2 - 4x})\right\}(x^2 - 4x)^{-1/4}$$

(ii) $$L_n^{(\alpha)}((n + \tfrac{1}{2})x) \sim \exp\left(\frac{x + \sqrt{x^2 - 4x}}{4}\right) L_n^{(\alpha)}(nx)$$

(iii) $$L_n^{(\alpha)}((n+1)x) \sim \exp\left(\frac{x + \sqrt{x^2 - 4x}}{2}\right) L_n^{(\alpha)}(nx) .$$

<u>Proof</u> : (i) Let $-x = p^2/(1-p)$ $(0 < p < 1)$ then from (0.1.9) we find

$$(3.2.3) \qquad L_n^{(\alpha)}(nx) = (1-p)^{-n} \sum_{j=0}^{n} \binom{n+\alpha}{n-j} \frac{n^j}{j!} p^{2j}(1-p)^{n-j} .$$

Consider two independent sequences $\{X_i : i = 1,2,...\}$ and $\{Y_i : i = 1,2,...\}$ where X_i are independent Bernoulli random variables with parameter p and Y_i are independent Poisson random variables with parameter p. If we put $S_n = X_1 + ... + X_n$ and $S_n^* = Y_1 + Y_2 + ... + Y_n$ then

$$P(S_{n+\alpha} - S_n^* = \alpha) = \sum_{j=0}^{n} P(S_{n+\alpha} = j + \alpha)P(S_n^* = j)$$

$$= e^{-np} p^{\alpha} \sum_{j=0}^{n} \binom{n+\alpha}{n-j} \frac{n^j}{j!} p^{2j} (1-p)^{n-j}$$

so that from (3.2.3) we obtain

$$(3.2.4) \qquad L_n^{(\alpha)}(nx) = \frac{e^{np} p^{-\alpha}}{(1-p)^n} P(S_{n+\alpha} - S_n^* = \alpha) .$$

Now $S_{n+\alpha} - S_n^*$ has the same distribution as $\sum_{j=1}^{n+\alpha} Z_j$, with

$$Z_j = \begin{cases} X_j & 1 \le j \le \alpha \\ X_j - Y_j & \alpha+1 \le j \le n+\alpha . \end{cases}$$

The sequence $\{Z_j : j = 1,2,...\}$ satisfies (3.1.2) and

$$m_{n+\alpha} = E(S_{n+\alpha} - S_n^*) = \alpha p$$

$$s_{n+\alpha}^2 = Var(S_{n+\alpha} - S_n^*) = np(2-p) + \alpha p(1-p)$$

so that the local limit theorem implies

$$\lim_{n \to \infty} \left| \sqrt{np(2-p) + \alpha p(1-p)}\, P(S_{n+\alpha} - S_n^* = \alpha) - \frac{1}{\sqrt{2\pi}} \exp\left\{ - \frac{\alpha^2(1-p)^2}{2(np(2-p) + \alpha p(1-p))} \right\} \right| = 0$$

from which

$$P(S_{n+\alpha} - S_n^* = \alpha) \sim \frac{1}{\sqrt{2\pi np(2-p)}}$$

follows. Insert this into (3.2.4) then the result follows by using the formulas

$$p = \frac{x + \sqrt{x^2 - 4x}}{2} \quad ; \quad \frac{1}{p} = \frac{-x + \sqrt{x^2 - 4x}}{-2x}$$

$$\frac{1}{1-p} = -\frac{1}{x}\left(\frac{-x + \sqrt{x^2 - 4x}}{2}\right)^2 \quad ; \quad 2-p = \frac{2\sqrt{x^2 - 4x}}{-x + \sqrt{x^2 - 4x}} .$$

(ii) Let $-x = p^2/(1-p)$ then

$$(3.2.5) \qquad L_n^{(\alpha)}((n + \tfrac{1}{2})x) = (1 - p)^{-n} \sum_{j=0}^{n} \binom{n+\alpha}{n-j} \frac{(2n+1)^j}{j!} \frac{p^{2j}}{2^j} (1 - p)^{n-j} .$$

Let $\{Y_i' : i = 1,2,\dots\}$ be a sequence of independent Poisson random variables with parameter $p/2$, independent of the sequence $\{X_i : i = 1,2,\dots\}$ of Bernoulli random variables with parameter p, then with $S_n' = Y_1' + \dots + Y_n'$ we have

$$P(S_{n+\alpha} - S_{2n+1}' = \alpha) = p^\alpha e^{-(n+1/2)p} \sum_{j=0}^{n} \binom{n+\alpha}{n-j} \frac{p^{2j}}{j!} \frac{(2n+1)^j}{2^j} (1-p)^{n-j}$$

so that comparison with (3.2.5) leads to

$$(3.2.6) \qquad L_n^{(\alpha)}((n + 1/2)x) = \frac{p^{-\alpha} e^{(n+1/2)p}}{(1 - p)^n} P(S_{n+\alpha} - S_{2n+1}' = \alpha) .$$

As in (i) we find

$$P(S_{n+\alpha} - S_{2n+1}' = \alpha) \sim \frac{1}{\sqrt{2\pi np(2-p)}} \sim P(S_{n+\alpha} - S_n^* = \alpha)$$

which gives the desired result.

(iii) Once more we let $x = -p^2/(1-p)$ then with the same notation as in (i).

$$P(S_{n+\alpha} - S_{n+1}^* = \alpha) = p^\alpha e^{-(n+1)p} \sum_{j=0}^{n} \binom{n+\alpha}{n-j} \frac{p^{2j}}{j!} (1-p)^{n-j} (n+1)^j$$

from which we obtain

$$L_n^{(\alpha)}((n+1)x) = \frac{p^{-\alpha}e^{(n+1)p}}{(1-p)^n} P(S^*_{n+\alpha} - S^*_{n+1} = \alpha) .$$

The local limit theorem gives

$$P(S_{n+\alpha} - S^*_{n+1} = \alpha) \sim \frac{1}{\sqrt{2\pi np(2-p)}} \sim P(S_{n+\alpha} - S^*_n = \alpha)$$

which gives the desired result.

■

We can use this result for (generalized) Laguerre polynomials to find a result for (generalized) Hermite polynomials $\{H_n^{(\alpha)}(x) : n = 0,1,2,\ldots\}$ defined in (0.1.14). The result corresponds to the formula of Plancherel-Rotach [156] on the imaginary axis (see also Szegö [175], Thm. 8.22.9b).

Theorem 3.5. : Suppose that $\alpha - 1/2$ is a positive integer and $y > 0$ then

$$H_n^{(\alpha)}(i\sqrt{2n}\, y) \sim \frac{\Gamma(\frac{n+2}{2})}{\sqrt{n\pi}} 2^{n-\alpha-1/2} (y + \sqrt{y^2+1})^{n+\alpha+1/2} y^{-\alpha}$$

$$\times \exp\left\{ny(-y + \sqrt{y^2+1})\right\} (y^2+1)^{-1/4} .$$

Proof : This follows immediately by using Theorem 3.4(i) and (ii) and the relation (0.1.14) between Hermite and Laguerre polynomials.

■

The results in Theorems 3.3 - 3.5 have been known for a long time and the probabilistic method only gives a simpler proof for these asymptotic formulas. We will now handle two other families of orthogonal polynomials for which an asymptotic formula was not known, so that the probabilistic approach gives an important insight for the asymptotic behaviour for these polynomials. First we treat Charlier polynomials $\{C_n^{(a)}(x) : n = 0,1,2,\ldots\}$ defined in (0.1.17).

Theorem 3.6. (Maejima-Van Assche [111]). Suppose that $a > 0$ and $x < 0$ then

(i) $C_n^{(a)}(nx) \sim \frac{(-1)^n}{\sqrt{2\pi n}} n! \, (1-x)^n \left(\frac{x-1}{x}\right)^{-nx-1/2} \exp\left(\frac{a}{1-x}\right)$

(ii) $C_n^{(a)}((n+1)x) \sim \left(\frac{x-1}{x}\right)^{-x} C_n^{(a)}(nx)$.

Proof : (i) Let $\{X_i : i = 1,2,\ldots\}$ be a sequence of independent random variables, all with the same negative-binomial distribution with parameters $\alpha = -x$ and $p = \frac{x}{x-1}$. If Y is a Poisson random variable with parameter $\frac{a}{1-x}$, independent of every X_i, then with $S_n = X_1 + X_2 + \ldots + X_n$

$$P(S_n + Y = n) = \sum_{j=0}^{n} P(S_n = j)P(Y = n - j)$$

$$= \frac{\exp(\frac{a}{x-1})(\frac{x}{x-1})^{-nx}}{(1-x)^n} \sum_{j=0}^{n} (-1)^j \binom{nx}{j} \frac{a^{n-j}}{(n-j)!} .$$

Comparison with (0.1.17) leads to

$$c_n^{(\alpha)}(nx) = (-1)^n\, n!\, \exp(\tfrac{a}{1-x})(\tfrac{x-1}{x})^{-nx} (1-x)^n\, P(S_n + Y = n) . \tag{3.2.7}$$

Clearly $S_n + Y$ has the same distribution as $\sum_{j=1}^{n+1} Z_j$, with

$$\begin{cases} Z_1 = Y \\ Z_j = X_j & 2 < j < n+1 . \end{cases}$$

The sequence $\{Z_i : i = 1,2,\ldots\}$ satisfies (3.1.2) and

$$m_{n+1} = E(S_n + Y) = n + \frac{a}{1-x}$$

$$s_{n+1}^2 = \mathrm{Var}(S_n + Y) = n\,\frac{x-1}{x} + \frac{a}{1-x}$$

so that the local limit theorem yields

$$\lim_{n \to \infty} \left| \sqrt{n\,\frac{x-1}{x} + \frac{a}{1-x}}\; P(S_n + Y = n) - \frac{1}{\sqrt{2\pi}} \exp\left\{- \frac{(\frac{a}{1-x})^2}{2(n\,\frac{x-1}{x} + \frac{a}{1-x})}\right\}\right| = 0$$

from which

$$P(S_n + Y = n) \sim \frac{1}{\sqrt{2\pi n\,\frac{x-1}{x}}}$$

follows. Insert this into (3.2.7) then the result follows.

(ii) With the same notation as in (i) one easily finds

$$c_n^{(a)}((n+1)x) = (-1)^n\ n!\ \exp(\tfrac{a}{1-x})(\tfrac{x-1}{x})^{-(n+1)x}(1-x)^n\ P(S_{n+1} + Y = n)\ .$$

The local limit theorem gives

$$P(S_{n+1} + Y = n) \sim \frac{1}{\sqrt{2\pi n\ \frac{x-1}{x}}} \sim P(S_n + Y = n)$$

which gives the desired result. ∎

Finally we handle the Meixner polynomials (of the first kind) $\{m_n(x;\beta,c) : n = 0,1,2,\ldots\}$ given in (0.1.20).

Theorem 3.7. (Maejima-Van Assche [111]). Let β be a positive integer and $x < 0$ then

(i) $$m_n(x;\beta,c) \sim \frac{2^{-2n-\beta}c^{-n}n!}{\sqrt{2\pi n}}\{-(1-c)(1+x) + \phi(x)\}^{n+\beta}$$

$$\times\ \{(1-c)(1-x) + \phi(x)\}^n\left\{1 + \frac{(1-c)(1+x) + \phi(x)}{-2x}\right\}^{-nx}$$

$$\times\ \{-(1-c)x\}^{-n-\beta+1/2}\phi(x)^{-1/2}\ ,$$

(ii) $$m_n((n+1)x;\beta,c) \sim \left\{1 + \frac{(1-c)(1+x) + \phi(x)}{-2x}\right\}^{-x} m_n(nx;\beta,c)$$

where

$$\phi(x) = \sqrt{(1-c)^2(1+x)^2 - 4x(1-c)}\ .$$

Proof : (i) Let $\{X_i : i = 1,2,\ldots\}$ be a sequence of independent Bernoulli random variables with parameter p and $\{Y_i : i = 1,2,\ldots\}$ a sequence of independent random variables having a negative-binomial distribution with parameters α and q, independent of every X_i. Taking $\alpha = -x$ then with $S_n = X_1 + \ldots + X_n$ and $S_n^* = Y_1 + \ldots + Y_n$ we have

$$(3.2.8) \qquad P(S_{n+\beta-1} - S_n^* = \beta-1) = \sum_{j=0}^{n} P(S_{n+\beta-1} = j+\beta-1)P(S_n^* = j)$$

$$= q^{-nx}(1-p)^n p^{\beta-1} \sum_{j=0}^{n} (-1)^j \binom{nx}{j}\binom{n+\beta-1}{j+\beta-1} p^j (1-q)^j (1-p)^{-j}$$

Let

$$p = \frac{1}{2}\left\{(1-c)(1+x) + \sqrt{(1-c)^2(1+x)^2 - 4x(1-c)}\right\}$$

then $1-c < p < 1$ whenever $x < 0$. Also define

$$q = \frac{-x}{p-x}$$

then $0 < q < 1$ for $x < 0$. Notice that

$$p + x\,\frac{1-q}{q} = 0 \quad , \qquad \frac{p}{1-p}\,(1-q) = \frac{1}{c} - 1.$$

Compare (3.2.8) with (0.1.20) then

$$(3.2.9) \qquad m_n(nx;\beta,c) = n!\; q^{nx}(1-p)^{-n}p^{1-\beta}P(S_{n+\beta-1} - S_n^* = \beta-1) .$$

Clearly $S_{n+\beta-1} - S_n^*$ has the same distribution as $\sum_{j=1}^{n+\beta-1} Z_j$, with

$$Z_j = \begin{cases} X_j & 1 \leq j \leq \beta-1 \\ X_j - Y_j & \beta \leq j \leq n+\beta-1 \end{cases}$$

so that the local limit theorem applies with

$$m_{n+\beta-1} = E(S_{n+\beta-1} - S_n^*) = (\beta-1)p$$

$$s_{n+\beta-1}^2 = Var(S_{n+\beta-1} - S_n^*) = np(2 - p\,\frac{1+x}{x}) + (\beta-1)p(1-p)$$

and implies

$$\lim_{n \to \infty} \Big| s_{n-\beta-1} \, P(S_{n+\beta-1} - S_n^* = \beta-1)$$

$$- \frac{1}{\sqrt{2\pi}} \exp\left\{- \frac{(\beta-1)^2(1-p)^2}{2(\beta-1)p(1-p) + 2np(2 - p\frac{1+x}{x})}\right\}\Big| = 0$$

which in turn leads to

$$P(S_{n+\beta-1} - S_n^* = \beta-1) \qquad \frac{1}{\sqrt{2\pi np(2 - p\,\frac{1+x}{x})}} \quad .$$

Use this in (3.2.9) then together with the formulas

$$\frac{1}{q} = 1 - \frac{p}{x} \quad , \qquad 2 - p\,\frac{1+x}{x} = \frac{2\phi(x)}{-(1-c)(1+x) + \phi(x)}$$

$$\frac{1}{1-p} = \frac{\{-(1-c)(1+x) + \phi(x)\}\{(1-c)(1-x) + \phi(x)\}}{-4c(1-c)x}$$

$$\frac{1}{p} = \frac{-(1-c)(1+x) + \phi(x)}{-2x(1-c)}$$

the result follows.

(ii) With the same notation as in (i) one easily finds

$$m_n((n+1)x;\beta,c) = n!\; q^{(n+1)x}\,(1-p)^n\,p^{1-\beta}\,P(S_{n+\beta-1} - S_{n+1}^* = \beta-1)$$

and by the local limit theorem

$$P(S_{n+\beta-1} - S_{n+1}^* = \beta-1) \sim P(S_{n+\beta-1} - S_n^* = \beta-1)$$

from which the result follows.

∎

Notice that when we take $y = (1-c)x$ in Theorem 3.7 and let c tend to one then we find the formulas in Theorem 3.4 with $\alpha = \beta-1$. This was to be expected since

$$\lim_{c \to 1} m_n\!\left(\frac{y}{1-c}\,;\beta,c\right) = L_n^{(\beta-1)}(y) \quad .$$

3.3. Asymptotic expansions

In this section we will use the asymptotic expansion for the local limit theorem (Theorem 3.2) to improve the results in the preceding section. We begin with the Gegenbauer polynomials $\{P_n^{(\alpha,\alpha)}(x) : n = 0,1,2,\ldots\}$.

Theorem 3.8. (Maejima-Van Assche [111]) : Let α be a positive integer and $x > 1$ then for every integer $M > 0$

$$P_n^{(\alpha,\alpha)}(x) = \frac{1}{\sqrt{2\pi(n+\alpha)}} (x^2-1)^{-\frac{\alpha}{2}-\frac{1}{4}} (\sqrt{x+1}+\sqrt{x-1})^{2\alpha}$$

$$\times \left(x+\sqrt{x^2-1}\right)^{n+1/2} e^{-\xi^2/2} \left\{1 + \sum_{\nu=0}^{M} \frac{f_{2\nu}(x)}{n^\nu} + o(n^{-M})\right\}$$

where

$$\xi = \xi(n,x) = \alpha\{(n+\alpha)\eta(x)\}^{-1/2} ,$$

$$\eta(x) = \frac{\sqrt{x^2-1}}{x+\sqrt{x^2-1}} ,$$

$$f_{2\nu}(x) = \sum{}^{*} H_{2\nu+2s}(\xi) \prod_{m=1}^{\nu} \frac{1}{k_m!} \left\{\frac{2K_{2m+2}(p)}{(2m+2)!\ \eta(x)^{m+2}}\right\}^{k_m} ,$$

$$p = \frac{\sqrt{x+1}}{\sqrt{x+1}+\sqrt{x-1}} ,$$

the sum $\sum^{*}$ in the expression for $f_{2\nu}$ runs over all non-negative integer solutions $(k_1,k_2,\ldots,k_\nu)$ of the equations $k_1 + 2k_2 + \ldots + \nu k_\nu = \nu$ and $k_1+k_2+\ldots+k_\nu = s$, H_m is the modified Hermite polynomial of degree m defined in (3.1.4) and $K_m(p)$ is given by (3.1.5).

Proof : As in (3.2.2) we have

$$P_n^{(\alpha,\alpha)}(x) = \frac{p^{-\alpha}(1-p)^{-2n-\alpha}}{(t^2-1)^n} P\left(\sum_{j=1}^{n+\alpha} Z_j = \alpha\right)$$

where $Z_j = X_j - Y_j$ and $\{X_i : i = 1,2,\ldots\}$ and $\{Y_i : i = 1,2,\ldots\}$ are independent Bernoulli random variables with parameter p. Notice that $E|Z_1|^k < \infty$ for every integer $k > 0$, $E(Z_1) = 0$ and $Var(Z_1) = \eta(x)$. Therefore we can use the asymptotic expansion in Theorem 3.2 to find

$$P_n^{(\alpha,\alpha)}(x) = \frac{p^{-\alpha}(1-p)^{-2n-\alpha}}{(t^2-1)^n} \frac{e^{-y^2/2}}{2\sqrt{\pi(n+\alpha)p(1-p)}} \left\{1 + \sum_{\nu=1}^{2M} \frac{g_\nu(y)}{n^{\nu/2}} + o(n^{-M})\right\}$$

with $y = \alpha\{2(n+\alpha)p(1-p)\}^{-1/2} = \xi(n,x)$. Furthermore

$$\gamma_m(Z_1) = \gamma_m(X_1) + (-1)^m \gamma_m(Y_1)$$

$$= \begin{cases} 0 & m \text{ odd} \\ 2K_m(p) & m \text{ even} . \end{cases}$$

If ν is odd then every non-negative integer solution $(h_1,h_2,\ldots,h_\nu)$ of the equations $h_1 + 2h_2 + \ldots + \nu h_\nu = \nu$ and $h_1 + h_2 + \ldots + h_\nu = s$ has $h_j \neq 0$ for some odd integer j so that

$$\prod_{j=1}^{\nu} \frac{1}{(h_j)!} \left\{\frac{K_{j+2}(Z_1)}{(j+2)!\sigma^{j+2}}\right\}^{h_j} = 0$$

and $g_\nu(y) = 0$ when ν is odd.

∎

The asymptotic expansion in the previous theorem corresponds to the well known expansion (Szegö [175], Theorems 8.21.3 and 8.21.10) but has a different form. Next we give an asymptotic expansion for Laguerre polynomials $\{L_n^{(0)}(x) : n = 0,1,2,\ldots\}$ equivalent with the expansion given by Moecklin [135] but also different in appearance :

<u>Theorem 3.9.</u> (Maejima-Van Assche [111]) : Suppose $x < 0$ then for every integer $M > 0$

$$L_n^{(0)}(nx) = \frac{1}{\sqrt{2\pi n}} \left(\frac{-x + \sqrt{x^2-4x}}{2}\right)^{2n+1} (-x)^{-n-1/2}$$

$$\times \exp\left\{\frac{n}{2}(x + \sqrt{x^2-4x})\right\}(x^2 - 4x)^{-1/4} \left\{1 + \sum_{\nu=1}^{M} \frac{d_{2\nu}(x)}{n^\nu} + o(n^{-M})\right\}$$

with

$$d_{2\nu} = \sum{}^{*} \frac{(2\nu + 2s)!}{(\nu+s)!} (-1)^{\nu+s}\, 2^{-\nu-s} \prod_{j=1}^{2\nu} \frac{1}{(h_j)!} \left\{ \frac{K_{j+2}(p) + (-1)^j p}{(j+2)![p(2-p)]^{1+j/2}} \right\}^{h_j} ,$$

$$p = \frac{x + \sqrt{x^2-4x}}{2} ,$$

the sum $\sum^{*}$ runs over all non-negative integer solutions $\{h_1,\ldots,h_\nu\}$ of the equations $h_1 + 2h_2 + \ldots + 2\nu h_{2\nu} = 2\nu$ and $h_1 + h_2 + \ldots + h_{2\nu} = s$ and $K_m(p)$ is given in (3.1.5).

Proof : As in (3.2.4) we have

$$L_n^{(0)}(nx) = (1-p)^{-n} e^{np} P\left(\sum_{j=1}^{n} Z_j = 0\right)$$

where $Z_j = X_j - Y_j$, with $\{X_i : i = 1,2,\ldots\}$ independent Bernoulli random variables with parameter p and $\{Y_i : i = 1,2,\ldots\}$ independent Poisson random variables with parameter p, independent of every X_i. Notice that $E|Z_1|^k < \infty$ for every integer $k > 0$, $E(Z_1) = 0$ and $Var(Z_1) = p(2-p)$. The asymptotic expansion in Theorem 3.2 then gives

$$L_n^{(0)}(nx) = \frac{1}{\sqrt{2\pi n}} \left(\frac{-x + \sqrt{x^2-4x}}{2}\right)^{2n+1} (-x)^{-n-1/2} \exp\left\{\frac{n}{2}\,(x+\sqrt{x^2-4x})\right\} (x^2-4x)^{-1/4}$$

$$\times \left\{1 + \sum_{\nu=1}^{2M} \frac{g_\nu(0)}{n^{\nu/2}} + o(n^{-M})\right\} .$$

We only need to consider even integers ν since $H_{\nu+2s}(0) = 0$ for odd ν. One easily verifies

$$H_{2\nu+2s}(0) = \frac{(2\nu+2s)!}{(\nu+s)!} (-1)^{\nu+s}\, 2^{-\nu-s} .$$

Furthermore

$$\gamma_m(Z_1) = \gamma_m(X_1) + (-1)^m \gamma_m(Y_1)$$

$$= K_m(p) + (-1)^m p$$

which gives the result.

∎

Finally we give an asymptotic expansion for the Meixner polynomials $\{m_n(nx;1,c) : n = 0,1,2,\ldots\}$:

<u>Theorem 3.10.</u> (Maejima-Van Assche [111]) : Suppose that $x < 0$ then for every integer $M > 0$

$$m_n(nx;1,c) = \frac{2^{-2n}c^{-n}n!}{\sqrt{2\pi n\xi(x)}}\left\{1 + \frac{(1-c)(1+x) + \phi(x)}{-2x}\right\}^{-nx}$$

$$\times\left\{-(1-c)(1+x) + \phi(x)\right\}^n\left\{(1-c)(1-x) + \phi(x)\right\}^n$$

$$\times\{-(1-c)x\}^{-n}\left\{1 + \sum_{\nu=1}^{M}\frac{f_{2\nu}(x)}{n^{\nu}} + o(n^{-M})\right\}$$

with

$$\phi(x) = \sqrt{(1-c)^2(1+x)^2 - 4x(1-c)}$$

$$\xi(x) = \frac{\{(1-c)(1+x) + \phi(x)\}\phi(x)}{-(1-c)(1+x) + \phi(x)}$$

$$f_{2\nu}(x) = \sum^{*}\frac{(2\nu+2s)!}{(\nu+s)!}(-1)^{\nu+s}2^{-\nu-s}\prod_{j=1}^{2\nu}\frac{1}{(h_j)!}\left\{\frac{K_{j+2}(p) + (-1)^j K^{*}_{j+2}(-x,q)}{(j+2)!\,\xi(x)^{j+2}}\right\}^{h_j},$$

the sum $\sum^{*}$ runs over all non-negative integer solutions $(h_1,\ldots,h_{2\nu})$ of the equations $h_1 + 2h_2 + \ldots + 2\nu h_{2\nu} = 2\nu$ and $h_1 + h_2 + \ldots + h_{2\nu} = s$, $K_n(p)$ is given in (3.1.5) and $K^{*}_n(-x,q)$ in (3.1.6), while

$$p = \frac{1}{2}\{(1-c)(1+x) + \phi(x)\} \quad ; \qquad q = \frac{-x}{p-x}.$$

<u>Proof</u> : As in (3.2.9) we have

$$m_n(nx;1,c) = q^{nx}(1-p)^{-n}\, n!\, P\Big(\sum_{j=1}^{n} Z_j = 0\Big)$$

where $Z_j = X_j - Y_j$, with $\{X_i : i = 1,2,\ldots\}$ independent Bernoulli random variables with parameter p and $\{Y_i : i = 1,2,\ldots\}$ independent negative-binomial random variables with parameters $-x$ and q, independent of every X_i. Notice that $E|Z_1|^k < \infty$ for every $k > 0$, $E(Z_1) = 0$ and $\mathrm{Var}(Z_1) = \xi(x)$. The asymptotic expansion (3.1.3) gives

$$m_n(nx;1,c) = q^{nx}(1-p)^{-n} \frac{1}{\sqrt{2\pi n\xi(x)}} \left\{1 + \sum_{\nu=1}^{2M} \frac{g_\nu(0)}{n^{\nu/2}} + o(n^{-M})\right\}$$

where

$$g_\nu(0) = \sum{}^{*} H_{\nu+2s}(0) \prod_{j=1}^{\nu} \frac{1}{(h_j)!} \left\{\frac{\gamma_{j+2}(Z_1)}{(j+2)!\xi(x)^{j+2}}\right\}^{h_j} .$$

The result then follows since

$$H_{\nu+2s}(0) = 0 \qquad \nu \text{ odd} \quad , \quad H_{2\nu+2s}(0) = \frac{(2\nu+2s)!}{(\nu+s)!} (-1)^{\nu+s} 2^{-\nu-s} ,$$

$$\gamma_m(Z_1) = K_m(p) + (-1)^m K_m^*(-x,q) .$$

■

These probabilistic proofs have the disadvantage that there does not seem to be a general rule for finding out which random variables correspond to a particular family of polynomials. Furthermore, the formulas are only valid for a restricted number of x-values due to the fact that probabilities are necessarily positive. There is also a restriction on some of the parameters in the families considered which has to do with the fact that we deal with lattice distributions. The probabilistic approach however does give precise information about the possible asymptotic behaviour of a sequence of polynomials. Notice that we never explicitely used the orthogonality of the polynomials in consideration. Methods of analytic continuation (e.g. the Stieltjes-Vitali theorem) may be used to get rid of the restrictions on the x variable in some cases (see [188] for Hermite polynomials).

CHAPTER 4 . ORTHOGONAL POLYNOMIALS ON UNBOUNDED SETS

In Chapters 1 and 2 we considered orthogonal polynomials on compact sets. In Chapter 3 we also studied Laguerre and Hermite polynomials which have weight functions on $[0,\infty)$ and $(-\infty,\infty)$ respectively. We also treated Charlier polynomials and Meixner polynomials which have a spectral measure on the positive integers. These sets are unbounded and the zeros can spread out to infinity. This flow to infinity can be handled in some cases. A first method consists in finding a linear mapping that maps the smallest zero $x_{1,n}$ to -1 and the largest zero $x_{n,n}$ to 1. The images $\{y_{j,n} : j = 1,2,\ldots,n\}$ of the zeros $\{x_{j,n} : j = 1,2,\ldots,n\}$ are called *contracted zeros* and by construction these are all contained in the interval $[-1,1]$. A second method of handling the zeros that tend to infinity consists of giving every zero $x_{j,n}$ a weight $\beta_{j,n}$ (different from $1/n$) in such a way that large zeros receive a small weight. This method leads to *weighted zero distributions*.

4.1. Contracted zero distributions

The asymptotic behaviour of orthogonal polynomials on a compact set with positive capacity is very closely related with the Frostman measure of that compact set (Chapter 1). The arcsin measure in particular is very important for orthogonal polynomials on $[-1,1]$. There is another measure on $[-1,1]$, actually a whole family of measures, that will play a similar role when studying the zeros of orthogonal polynomials on an infinite interval :

Definition : The *Ullman measure* μ^α with parameter $\alpha > 0$ is defined by

$$(4.1.1) \qquad \mu^\alpha(A) = \int_A v(\alpha;t)\,dt$$

where A is a Borel set in $[-1,1]$ and $v(\alpha;t)$ is the weight function given by

$$(4.1.2) \qquad v(\alpha;t) = \frac{\alpha}{\pi}\int_{|t|}^{1} \frac{y^{\alpha-1}}{\sqrt{y^2-t^2}}\,dy \qquad -1 < t < 1 .$$

This measure was named after Joseph Ullman who introduced it for the parameters

$\alpha = 2,4,6$ in [185]. The potential of the Ullman measure μ^α has the property

$$(4.1.3) \qquad U(x;\mu^\alpha) \geq -\frac{|x|^\alpha}{\lambda_\alpha} + \log 2 + \frac{1}{\alpha} \qquad x \in \mathbb{R}$$

and equality holds for $x \in [-1,1]$. Here

$$(4.1.4) \qquad \lambda_\alpha = \frac{2}{\sqrt{\pi}} \frac{\Gamma(\frac{\alpha+1}{2})}{\Gamma(\frac{\alpha}{2})} = \frac{\alpha}{\pi} \int_{-1}^{1} \frac{|x|^\alpha}{\sqrt{1-x^2}}\, dx .$$

The Ullman measure μ^α is the unique measure that minimizes the energy in the "external field" $|x|^\alpha/\lambda_\alpha$ in the sense that

$$(4.1.5) \qquad \min_{\mu \in \Omega} I(\mu;\alpha) = I(\mu^\alpha;\alpha)$$

where Ω is the set of all probability measures on $[-1,1]$ and

$$I(\mu;\alpha) = \int_{-1}^{1}\int_{-1}^{1} \log \frac{1}{|x-y|}\, d\mu(x)d\mu(y) + \int_{-1}^{1} \frac{|x|^\alpha}{\lambda_\alpha}\, d\mu(x)$$

(Mhaskar-Saff [133], Rakhmanov [163], Gonchar-Rakhmanov [81]). The Ullman measure has the following probabilistic interpretation :

Lemma 4.1. : Let X be a random variable having the arcsin distribution on $[-1,1]$ and Y be a random variable with a beta$(\alpha,1)$ distribution on $[0,1]$, i.e.

$$P(X \leq t) = \frac{1}{\pi} \int_{-1}^{t} \frac{1}{\sqrt{1-x^2}}\, dx \qquad -1 \leq t \leq 1$$

$$P(Y \leq t) = t^\alpha \qquad 0 \leq t \leq 1 .$$

If X and Y are independent then XY has the Ullman distribution with parameter α on $[-1,1]$:

$$P(XY \leq t) = \int_{-1}^{t} v(\alpha;x)dx \qquad -1 \leq t \leq 1 .$$

Proof : Use the law of total probability then for $t \in [-1,0]$

$$P(XY \leq t) = \int_{0}^{1} P(XY \leq t | Y = y) dP(Y \leq y)$$

$$= \int_0^1 P(X < \frac{t}{y}) dP(Y < y)$$

where we have used the independence. If $\frac{t}{y} < -1$ then $P(X < \frac{t}{y}) = 0$ and therefore

$$P(XY < t) = \int_{|t|}^1 P(X < \frac{t}{y}) dP(Y < y) .$$

Using the explicit expressions for the distribution functions of X and Y gives

$$P(XY < t) = \frac{1}{\pi} \int_{|t|}^1 \int_{-1}^{t/y} \frac{1}{\sqrt{1 - x^2}} dx \, dy^\alpha .$$

Setting $x = u/y$ (with y fixed) and changing the order of integration gives

$$P(XY < t) = \frac{1}{\pi} \int_{-1}^t \int_{|u|}^1 \frac{dy^\alpha}{\sqrt{y^2 - u^2}} du = \int_{-1}^t v(\alpha;u) du .$$

A similar calculation works for $t \in [0,1]$ also. ∎

The random variable X is distributed on [-1,1] according to the Frostman measure on [-1,1]. Therefore we can think of the Ullman measure μ^α as a perturbed Frostman measure where the random variable Y indicates the perturbation. If α tends to infinity then Y becomes degenerate at one and the Ullman measure μ^α converges weakly to the arcsin measure on [-1,1]. This probabilistic interpretation enables us to find the moments of the Ullman measure in a simple way. Since X and Y are independent we find for every positive integer M

$$(4.1.6) \qquad \int_{-1}^1 t^M v(\alpha;t) dt = E(X^M Y^M) = E(X^M) E(Y^M)$$
$$= \frac{\alpha}{M+\alpha} \frac{1}{\pi} \int_{-1}^1 \frac{t^M}{\sqrt{1 - t^2}} dt .$$

A sufficient condition for a sequence of probability measures to converge weakly to the Ullman measure μ^α is given by the following lemma, which is a modification of Lemma 1.1 :

<u>Lemma 4.2.</u> : Let $\{\mu_n : n = 1,2,...\}$ be a sequence of probability measures on $\mathbb{R}$ such that μ_n converges weakly to some probability measure μ on [-1,1]. If

$$(4.1.7) \qquad \liminf_{n \to \infty} U(x;\mu_n) + \frac{|x|^\alpha}{\lambda_\alpha} \geq \log 2 + \frac{1}{\alpha} \qquad x \in [-1,1] \backslash B$$

where B has Lebesgue measure zero, then μ is equal to the Ullman measure μ^{α}.

Proof : Let μ_E be the arcsin measure on $E = [-1,1]$ then from (4.1.7) we immediately find

$$(4.1.8) \qquad \log 2 + \frac{1}{\alpha} \le \int_{-1}^{1} \liminf_{n\to\infty} \left(U(x;\mu_n) + \frac{|x|^{\alpha}}{\lambda_{\alpha}}\right) d\mu_E(x) .$$

By Fubini's theorem we have

$$\int_{-1}^{1} U(x;\mu_n)d\mu_E(x) = \int U(x;\mu_E)d\mu_n(x) \le \log 2$$

the inequality follows from (1.1.6). Fatou's lemma and (4.1.4) yield

$$\int_{-1}^{1} \liminf_{n\to\infty} \left(U(x;\mu_n) + \frac{|x|^{\alpha}}{\lambda_{\alpha}}\right) d\mu_E(x)$$

$$\le \liminf_{n\to\infty} \int_{-1}^{1} U(x;\mu_n)d\mu_E(x) + \int_{-1}^{1} \frac{|x|^{\alpha}}{\lambda_{\alpha}}\, d\mu_E(x)$$

$$\le \log 2 + \frac{1}{\alpha} .$$

A combination of this with (4.1.8) gives

$$\liminf_{n\to\infty} U(x;\mu_n) + \frac{|x|^{\alpha}}{\lambda_{\alpha}} = \log 2 + \frac{1}{\alpha} \qquad x \in [-1,1]\backslash B_1$$

where $\mu_E(B_1) = 0$. The lower envelope theorem for potentials then leads to

$$U(x;\mu) + \frac{|x|^{\alpha}}{\lambda_{\alpha}} \le \log 2 + \frac{1}{\alpha} \qquad x \in [-1,1]\backslash B_1 .$$

The lower semicontinuity can be used to show that the inequality holds for every x in [-1,1] and by (4.1.5) it then follows that $\mu = \mu^{\alpha}$. ∎

The next theorem shows why the Ullman measures are important for orthogonal polynomials on infinite intervals. First some more notation. Introduce a sequence of measures $\{\nu_n : n = 1,2,...\}$ by

$$(4.1.9)\qquad \begin{cases} \nu_n\left(\left\{(\frac{\lambda_\alpha}{2n})^{1/\alpha} x_{j,n}\right\}\right) = \frac{1}{n} & j = 1,2,\ldots,n \\ \\ \nu_n(A) = 0 & A \text{ contains no zeros of } p_n((\frac{2n}{\lambda_\alpha})^{1/\alpha} x) \end{cases}$$

where λ_α is defined in (4.1.4) and $\{x_{j,n} : j = 1,\ldots,n;\ n = 1,2,\ldots\}$ are zeros of orthogonal polynomials $\{p_n(x) : n = 1,2,\ldots\}$. This sequence of measures describes contracted zeros of orthogonal polynomials with contraction factor $(\lambda_\alpha/2n)^{1/\alpha}$.

<u>Theorem 4.3.</u> : (Rakhmanov [163], Mhaskar-Saff [133]) Let μ be an absolutely continuous spectral measure on the real line with an even weight function w that is positive almost everywhere in $\mathbb{R}$ and for which

$$(4.1.10)\qquad \lim_{|x|\to\infty} \frac{\log w(x)}{|x|^\alpha} = -1 \qquad\qquad \alpha > 0\,,$$

then

$$(4.1.11)\qquad \lim_{n\to\infty} n^{-1/\alpha} k_n^{-1/n} = \frac{1}{2}\left(\frac{2}{e\lambda_\alpha}\right)^{1/\alpha}$$

and

$$(4.1.12)\qquad \nu_n \twoheadrightarrow \mu^\alpha$$

where k_n is the leading coefficient of $p_n(x;\mu)$ and μ^α is the Ullman measure with parameter α.

<u>Proof</u> : First we will show that

$$\limsup_{n\to\infty} n^{-1/\alpha} k_n^{-1/n} \le \frac{1}{2}\left(\frac{2}{e\lambda_\alpha}\right)^{1/\alpha}.$$

Let $\{y_{j,n} : j = 1,\ldots,n\}$ be an array of points in $[-1,1]$ such that the measures $\{\nu_n^* : n = 1,2,\ldots\}$ defined by

$$\begin{cases} \nu_n^*(\{y_{j,n}\}) = \frac{1}{n} & j = 1,2,\ldots,n \\ \\ \nu_n^*(A) = 0 & A \text{ contains no } y_{j,n}\ (j = 1,\ldots,n) \end{cases}$$

are such that $\nu_n \twoheadrightarrow \mu^\alpha$ and let $q_n(x) = \prod_{j=1}^{n} (x-y_{j,n})$. By the minimal property

(0.2.17) we have

$$(4.1.13)\qquad k_n^{-2} \le (cn^{1/\alpha})^{2n} \int_{-\infty}^{\infty} q_n^2\left(\frac{x}{cn^{1/\alpha}}\right) w(x)\,dx$$

$$= (cn^{1/\alpha})^{2n+1} \left\{ \int_{-r}^{r} q_n^2(x) w(cn^{1/\alpha}x)\,dx + \int_{|x|>r} q_n^2(x) w(cn^{1/\alpha}x)\,dx \right\}$$

where c and r will be determined later. Let $w(cn^{1/\alpha}x) = \exp(-2nh_n(x))$ then by (4.1.10)

$$(4.1.14)\qquad \lim_{n\to\infty} h_n(x) = \lim_{n\to\infty} -\frac{\log w(cn^{1/\alpha}x)}{c^\alpha n|x|^\alpha}\,\frac{c^\alpha|x|^\alpha}{2} = \frac{c^\alpha}{2}\,|x|^\alpha$$

so there exists an integer n_0 and a positive number r_0 such that

$$(4.1.15)\qquad h_n(x) \ge 2\log 2|x| \qquad\qquad |x| \ge r_0\,,\ n \ge n_0\,.$$

Choose $r \ge r_0$ such that $r > e^{1/\alpha}$ and choose $c = \left(\frac{2}{\lambda_\alpha}\right)^{1/\alpha}$ then by the lower envelope theorem and (4.1.3)

$$\liminf_{n\to\infty} U(x;\nu_n^*) \ge U(x;\mu^\alpha) \ge -\frac{|x|^\alpha}{\lambda_\alpha} + \log 2 + \frac{1}{\alpha} \qquad\qquad x \in \mathbb{R}$$

and by (4.1.14)

$$\liminf_{n\to\infty} \{U(x;\nu_n^*) + h_n(x)\} \ge \log 2 + \frac{1}{\alpha} \qquad\qquad x \in \mathbb{R}.$$

This means that for any $\varepsilon > 0$ we can find an integer $n_1(\varepsilon)$ such that

$$(4.1.16)\qquad U(x;\nu_n^*) + h_n(x) > \log 2 + \frac{1}{\alpha} - \varepsilon \qquad\qquad |x| \le r,\ n \ge n_1\,.$$

Notice that

$$U(x;\nu_n^*) + h_n(x) = -\frac{1}{2n}\log\{q_n^2(x) w(cn^{1/\alpha}x)\}$$

so that from (4.1.13) we obtain

$$k_n^{-2n} \le (cn^{1/\alpha})^{2n+1} \left\{ \int_{-r}^{r} \exp[-2n(U(x;\nu_n^*) + h_n(x))]\,dx \right.$$

$$\left. + \int_{|x|>r} \exp[-2n(U(x;\nu_n^*) + h_n(x))]\,dx \right\}.$$

Clearly $U(x;\nu_n^*) > -\log(2|x|)$ when $|x| > r$ and if we use this and (4.1.15) in the second integral and (4.1.16) in the first then for large enough n

$$k_n^{-2} < (cn^{1/\alpha})^{2n+1}\left\{2r2^{-2n}e^{-2n/\alpha}e^{2n\varepsilon} + \int_{|x|>r} (2|x|)^{-2n}\, dx\right\}$$

$$= (cn^{1/\alpha})^{2n+1}\left\{2r2^{-2n}e^{-2n/\alpha}e^{2n\varepsilon} + 2^{-2n+1}\frac{r^{-2n+1}}{2n-1}\right\}$$

recall that $r > e^{1/\alpha}$ and $c = (\frac{2}{\lambda_\alpha})^{1/\alpha}$ so that

$$\limsup_{n\to\infty} n^{-1/\alpha}k_n^{-1/n} < \frac{1}{2}\left(\frac{2}{e\lambda_\alpha}\right)^{1/\alpha}e^{\varepsilon}$$

and since this is true for every $\varepsilon > 0$ we have what we wanted to show. Suppose next that $n^{-1/\alpha}k_n^{-1/n}$ does not converge to $\frac{1}{2}(\frac{2}{e\lambda_\alpha})^{1/\alpha}$ then

$$\liminf_{n\to\infty} n^{-1/\alpha}k_n^{-1/n} < \beta < \frac{1}{2}\left(\frac{2}{e\lambda_\alpha}\right)^{1/\alpha}$$

and we will choose $\beta > 0$. This means that there exists an increasing sequence $\{m_n : n = 1,2,...\}$ such that

$$\lim_{n\to\infty}\left\{\int_{-cm_n^{1/\alpha}}^{cm_n^{1/\alpha}} |\hat{p}_{m_n}(x)|^2\, w(x)\, dx\right\}^{1/m_n} < \beta^2$$

or

$$\lim_{n\to\infty}\left\{\int_{-1}^{1} |\hat{p}_{m_n}(cm_n^{1/\alpha}x)|^2\, w(cm_n^{1/\alpha}x)dx\right\}^{1/m_n} < \beta^2 .$$

Use Lemma 1.3 with $f_n(x) = \beta^{-2n}|\hat{p}_n(cn^{1/\alpha}x)|^2\, w(cn^{1/\alpha}x)$ and the fact that $w(cn^{1/\alpha}x)^{1/2n}$ converges to $\exp(-|x|^\alpha/\lambda_\alpha)$ to find a subsequence $\{t_n : n = 1,2,...\}$ such that

$$\limsup_{n\to\infty} |\hat{p}_{t_n}(ct_n^{1/\alpha}x)|^{1/t_n} \exp\left(-\frac{|x|^\alpha}{\lambda_\alpha}\right) < \beta \qquad x \in [-1,1]\backslash B_1$$

with B_1 a Borel set of Lebesgue measure zero. The remainder of the proof is exactly the same as the proof of Theorem 1.2 but now one uses Lemma 4.2 instead of Lemma 1.1.
■

This theorem was first given by Rakhmanov [163] for $\alpha > 1$ and by Mhaskar and Saff

[133] for the weight functions $w(x) = \exp(-|x|^{\alpha})$, with $\alpha > 0$. The present proof uses some ideas of Gonchar and Rakhmanov [81]. Ullman [185] was actually the first to prove something like Theorem 4.3 but this result only covered weight functions of the form $|x|^{\beta} \exp(-|x|^{\alpha})$ with $\beta > -1$ and $\alpha = 2,4,6$.

If we combine Theorem 4.3 with the probabilistic interpretation of the Ullman measure then we can obtain stochastic approximations for the zeros $\{x_{j,n} : j = 1,\dots,n\}$ as follows. Take 2n independent random variables $\{X_i : i = 1,\dots,n\}$ and $\{Y_i : i = 1,\dots,n\}$ where the X_i have the arcsin distribution on $[-1,1]$ and the Y_i have the beta$(\alpha,1)$ distribution on $[0,1]$. The sequence $\{Z_i = X_iY_i : i = 1,\dots,n\}$ then has the same asymptotic distribution as the contracted zeros $\{(\frac{\lambda\alpha}{2n})^{1/\alpha} x_{j,n} : j = 1,2,\dots,n\}$ so that $\left\{(\frac{2n}{\lambda\alpha})^{1/\alpha} Z_{j:n} : j = 1,2,\dots,n\right\}$ is a random approximation for $\{x_{j,n} : j = 1,2,\dots,n\}$, where $Z_{1:n} < Z_{2:n} < \dots < Z_{n:n}$ are the order statistics.

4.2. Freud conjectures

Consider the weight functions

$$(4.2.1) \qquad w(x) = |x|^{\beta} e^{-|x|^{\alpha}} \qquad \beta > -1 \ , \ \alpha > 0 .$$

These weight functions, and generalizations of these, are nowadays often referred to as *Freud weights*. Freud [63] - [67] formulated two conjectures for the weights (4.2.1). A first conjecture is about the leading coefficients $\{k_n : n = 1,2,\dots\}$ of the orthonormal polynomials $\{p_n(x) : n = 1,2,3,\dots\}$ corresponding to these weight functions, namely

$$(4.2.2) \qquad \lim_{n \to \infty} n^{-1/\alpha} \frac{k_{n-1}}{k_n} = \frac{1}{2} \left(\frac{2}{\lambda_\alpha}\right)^{1/\alpha} .$$

Notice that the ratio k_{n-1}/k_n is the coefficient a_n in the recurrence relation for the orthogonal polynomials and that (4.2.2) implies (4.1.11). A second conjecture involves the largest zero $x_{n,n}$ of the orthogonal polynomial $p_n(x)$:

$$(4.2.3) \qquad \lim_{n \to \infty} n^{-1/\alpha} x_{n,n} = \left(\frac{2}{\lambda_\alpha}\right)^{1/\alpha} .$$

Lemma 4.4. : If (4.2.2) holds then also (4.2.3) is true.

Proof : Notice that the weight function (4.2.1) is symmetric around zero. One can easily find lower and upper bounds for the largest zero $x_{n,n}$ in terms of the ratios k_{n-1}/k_n. A very useful one is (Freud [63])

$$(4.2.4) \qquad 2\,\frac{\sum_{j=1}^{n-1} \frac{k_{j-1}}{k_j} c_j c_{j+1}}{\sum_{j=1}^{n} c_j^2} < x_{n,n} < 2 \max_{1 < j < n-1} \frac{k_{j-1}}{k_j}$$

where $\{c_j : j = 1,\ldots,n\}$ is an arbitrary sequence of real numbers. The upper bound in (4.2.4) and (4.2.2) immediately lead to

$$\limsup_{n\to\infty} n^{-1/\alpha} x_{n,n} < \left(\frac{2}{\lambda_\alpha}\right)^{1/\alpha} .$$

On the other hand consider the sequence $c_j = j^m$ $(m > 0)$. If $\{a_n : n = 1,2,\ldots\}$ is a sequence such that $a_n \longrightarrow a$ $(n \to \infty)$ then

$$\lim_{n\to\infty} \frac{1}{n^{m+1}} \sum_{j=1}^{n} j^m a_j = \frac{a}{\beta+1}$$

so that

$$2 \lim_{n\to\infty} n^{-1/\alpha} \frac{1}{\sum_{j=1}^{n} j^{2m}} \sum_{j=1}^{n-1} \frac{1}{j^{1/\alpha}} \frac{k_{j-1}}{k_j} j^{m+1/\alpha}(j+1)^m$$

$$= \frac{2m+1}{2m+\frac{1}{\alpha}+1} \left(\frac{2}{\lambda_\alpha}\right)^{1/\alpha} .$$

If we let m tend to infinity then

$$\liminf_{n\to\infty} n^{-1/\alpha} x_{n,n} > \left(\frac{2}{\lambda_\alpha}\right)^{1/\alpha}$$

which gives (4.2.3).

∎

Freud [65] proved the conjecture (4.2.2) for $\alpha = 2,4$ and 6 but his method could not be extended to other values of α. Rakhmanov [163] was able to prove (4.2.3) for weight functions that are positive almost everywhere on $\mathbb{R}$ and for which (4.1.10) holds with $\alpha > 1$. Magnus [112] - [113] succeeded to prove (4.2.2) for every even integer α. Freud's conjecture (4.2.2) has been proved for a very general class of weight functions by Lubinsky-Mhaskar-Saff [108] - [109]. Their result is

<u>Theorem 4.5.</u> (Lubinsky-Mhaskar-Saff [108] - [109]). Suppose $w(x) = |x|^\rho \exp(-Q(x))$ where $\rho > -1$, Q is even and continuous and Q' exists for $x > 0$, while $xQ'(x)$

remains bounded as x tends to zero. Further assume that Q''' exists for x large enough and that for some $\alpha > 0$ and $c > 0$

$$\left.\begin{array}{l} Q'(x) > 0 \quad , \\ \\ \dfrac{x^2 |Q'''(x)|}{Q'(x)} < c \ , \end{array}\right| \qquad x \text{ large enough}$$

and

$$\lim_{x \to \infty} \left(1 + \frac{xQ''(x)}{Q'(x)}\right) = \alpha \ .$$

Then

$$\lim_{n \to \infty} \frac{1}{q_n} \frac{k_{n-1}}{k_n} = \frac{1}{2} \tag{4.2.5}$$

where q_n is the positive root of the equation

$$n = \frac{1}{\pi} \int_0^1 \frac{q_n x Q'(q_n x)}{\sqrt{1 - x^2}} \, dx \tag{4.2.6}$$

If $Q(x) = |x|^\alpha$ then one easily finds $q_n = (\frac{2n}{\lambda_\alpha})^{1/\alpha}$ which gives Freud's conjecture (4.2.2) for the weight functions (4.2.1). A proof of Theorem 4.5 needs some heavy tools of approximation theory and is therefore omitted. It is a big surprise that the same techniques needed to prove Theorem 4.5 lead to a much stronger result :

Theorem 4.6. (Lubinsky-Saff [110]). Suppose $w(x) = \exp(-Q(x))$ where Q satisfies the same conditions as in Theorem 4.5 then

$$\lim_{n \to \infty} \left(\frac{q_n}{2}\right)^{n+1/2} \exp\left\{ -\frac{1}{\pi} \int_0^1 \frac{Q(q_n t)}{\sqrt{1-t^2}} \, dt \right\} k_n = \sqrt{2\pi} \tag{4.2.7}$$

where q_n has the same meaning as in the previous theorem.

If one considers the weight function $e^{-|x|^\alpha}$ with α an even integer then one can give more precise results concerning (4.2.2) and (4.2.3). First of all one can give an asymptotic expansion for (4.2.2) :

Theorem 4.7. (Máté-Nevai-Zaslavsky [129]). Let α be an even positive integer then for the weight function $e^{-|x|^\alpha}$ one has

$$(4.2.8) \qquad n^{-1/\alpha}\frac{k_{n-1}}{k_n} = \sum_{j=0}^{\infty} c_j n^{-2j}$$

where $c_0 = \frac{1}{2}(\frac{2}{\lambda_\alpha})^{1/\alpha}$.

The coefficients c_j in this asymptotic expansion are not explicitely known. The proof of this result is based on the fact that $a_n = k_{n-1}/k_n$ satisfies a non-linear recurrence relation which is "smooth" enough so that it can be handled asymptotically.

The result in Theorem 4.7 can then be used to find precise asymptotic expressions for the largest zeros of orthogonal polynomials for weight function $e^{-|x|^\alpha}$ with α an even integer. Recall that *Airy's function* is the unique solution of the differential equation

$$y'' + \frac{xy}{3} = 0$$

which remains bounded as $x \longrightarrow -\infty$.

Theorem 4.8. (Máté-Nevai-Totik [124]-[125]). Let α be a positive even integer and let $i_1 < i_2 < i_3$ ... be the zeros of Airy's function, then the largest zeros of the orthogonal polynomials with weight function $e^{-|x|^\alpha}$ satisfy

$$(4.2.9) \qquad n^{-1/\alpha}x_{n-k+1,n} = (\frac{2}{\lambda_\alpha})^{1/\alpha}\left\{1 - \frac{i_k}{(6n^2\alpha^2)^{1/3}}\right\} + o(n^{-2/3})$$

for every fixed integer k.

For more information concerning the Freud conjectures I refer to Nevai's excellent survey [151].

4.3. Results using the recurrence relation

Nevai and Dehesa [152] have used the recurrence relation

$$xp_n(x) = a_{n+1}p_{n+1}(x) + b_np_n(x) + a_np_{n-1}(x) \qquad n = 0,1,2,...$$

with recurrence coefficients $\{a_n,b_n\}$ that tend to infinity in a particular way and were able to find some asymptotic average properties of the zeros $\{x_{j,n} : j = 1,2,...,n\}$:

Theorem 4.9 (Nevai-Dehesa [152]) : Let $\varphi : \mathbb{R}^+ \longrightarrow \mathbb{R}^+$ be a non-decreasing function such that for every $t \in R$

$$(4.3.1) \qquad \lim_{x \to \infty} \frac{\varphi(x+t)}{\varphi(x)} = 1$$

and suppose that there exist $a > 0$ and $b \in \mathbb{R}$ such that

$$(4.3.2) \qquad \lim_{n \to \infty} \frac{a_n}{\varphi(n)} = a \quad , \qquad \lim_{n \to \infty} \frac{b_n}{\varphi(n)} = b \; ,$$

then for every positive integer M

$$(4.3.3) \qquad \lim_{n \to \infty} \frac{1}{\int_0^n \varphi(t)^M \, dt} \sum_{j=1}^n x_{j,n}^M = \begin{cases} b^M & \text{if } a = 0 \\ \frac{1}{\pi} \int_{b-2a}^{b+2a} \frac{t^M}{\sqrt{4a^2 - (t-b)^2}} \, dt & \text{if } a > 0 \end{cases}$$

For the special case $b = 0$, $a = 1/2$ and $\varphi(t) = t^{1/\alpha}$ $(\alpha > 0)$ the expression in (4.3.3) becomes

$$(4.3.4) \qquad \lim_{n \to \infty} \frac{1}{n} \sum_{j=1}^n \left(\frac{x_{j,n}}{an^{1/\alpha}}\right)^M = \frac{\alpha}{M+\alpha} \frac{1}{\pi} \int_{-1}^{1} \frac{t^M}{\sqrt{1-t^2}} \, dt \; .$$

If we introduce the measures $\{\nu_n : n = 1,2,\ldots\}$ by

$$\begin{cases} \nu_n(\{\frac{x_{j,n}}{an^{1/\alpha}}\}) = \frac{1}{n} & j = 1,2,\ldots,n \\ \nu_n(A) = 0 & \text{if } A \text{ contains no zeros of } p_n(an^{1/\alpha} x) \end{cases}$$

then the left hand side in (4.3.4) consists of the Mth moment of ν_n and the right hand side is, after comparison with (4.1.6), the Mth moment of the Ullman measure μ^α. The support of the Ullman measure μ^α is compact and therefore the convergence of the moments implies the weak convergence of the measures ν_n to the Ullman measure μ^α. The Ullman measure thus enters in a natural way when the recurrence coefficients a_n tend to infinity as a power of n. This was to be expected since this is exactly the asymptotic behaviour predicted by Freud's conjecture (4.2.2).

The same conditions (4.3.1) and (4.3.2) can be used to find the asymptotic behaviour of the ratio $p_{n-1}(\varphi(n)x)/p_n(\varphi(n)x)$ in the complex plane :

Theorem 4.10. (Van Assche [191]) : Suppose that (4.3.1) and (4.3.2) are valid, then for every $a > 0$

$$\lim_{n\to\infty} \frac{p_{n-1}(\varphi(n)x)}{p_n(\varphi(n)x)} = \frac{2a}{x-b+\sqrt{(x-b)^2-4a^2}} \tag{4.3.5}$$

uniformly on compact sets of $\mathbb{C}\setminus[A,B]$, where $[A,B]$ is the smallest interval that contains $\{0\}$ and $[b-2a,b+2a]$. If $a = 0$ then

$$\lim_{n\to\infty} \frac{\varphi(n)}{a_n} \frac{p_{n-1}(\varphi(n)x)}{p_n(\varphi(n)x)} = \frac{1}{x-b} \tag{4.3.6}$$

uniformly on compact subsets of $\mathbb{C}\setminus[0,b]$ (if $b > 0$) or $\mathbb{C}\setminus[b,0]$ (if $b < 0$).

Proof : We will only consider the case where $a > 0$, when $a = 0$ a similar reasoning applies. Consider the Turán determinant

$$D_k(x) = p_k^2(x) - \frac{a_{k+1}}{a_k} p_{k+1}(x)p_{k-1}(x) , \tag{4.3.7}$$

then by the recurrence relation applied to $a_{k+1}p_{k+1}(x)$ we obtain

$$D_k(x) = p_k^2(x) - \frac{1}{a_k} p_{k-1}(x) \left\{(x-b_k)p_k(x) - a_k p_{k-1}(x)\right\}$$

$$= D_{k-1}(x) + p_k(x) \left\{\frac{a_k}{a_{k-1}} p_{k-2}(x) + p_k(x) - \frac{x-b_k}{a_k} p_{k-1}(x)\right\} .$$

Use the recurrence relation again for $p_k(x)$ between the curly brackets, then

$$D_k(x) = D_{k-1}(x) + \frac{b_k - b_{k-1}}{a_k} p_k(x)p_{k-1}(x) + \left(\frac{a_k}{a_{k-1}} - \frac{a_{k-1}}{a_k}\right)p_{k-2}(x)p_k(x) \tag{4.3.8}$$

Let

$$R_{k,n}(x) = \frac{D_k(\varphi(n)x)}{p_{k+1}(\varphi(n)x)p_k(\varphi(n)x)}$$

where $x \in [B+d,\infty)$ (so that the denominator does not vanish), then from (4.3.8) we get

$$(4.3.9) \qquad |R_{k,n}(x)| \leq |R_{k-1,n}(x)| \frac{|p_{k-1}(\varphi(n)x)|}{|p_{k+1}(\varphi(n)x)|}$$

$$+ \frac{|b_k - b_{k-1}|}{a_k} \frac{|p_{k-1}(\varphi(n)x)|}{|p_{k+1}(\varphi(n)x)|} + \left|\frac{a_k}{a_{k-1}} - \frac{a_{k-1}}{a_k}\right| \frac{|p_{k-2}(\varphi(n)x)|}{|p_{k+1}(\varphi(n)x)|}$$

Using (0.2.10) gives

$$\frac{|p_k(\varphi(n)x)|}{|p_{k+1}(\varphi(n)x)|} \leq \sum_{j=1}^{k+1} \frac{a_{j,k+1}}{|\varphi(n)x - x_{j,k+1}|} \leq \frac{1}{d\varphi(n)} \sum_{j=1}^{k+1} a_{j,k+1} = \frac{a_{k+1}}{d\varphi(n)}$$

and there exists a constant C such that $a_k/\varphi(n) \leq C$ for every $k \leq n$, so that

$$\frac{|p_k(\varphi(n)x)|}{|p_{k+1}(\varphi(n)x)|} \leq \frac{C}{d} \qquad\qquad x \in [B+d,\infty)$$

for every $k \leq n$ and n large enough. Therefore from (4.3.9) we get

$$|R_{k,n}(x)| \leq |R_{k-1,n}(x)| \left(\frac{C}{d}\right)^2 + A_k$$

where

$$A_k = \frac{|b_k - b_{k-1}|}{a_k} \left(\frac{C}{d}\right)^2 + \left|\frac{a_k}{a_{k-1}} - \frac{a_{k-1}}{a_k}\right| \left(\frac{C}{d}\right)^3 .$$

Iteration yields

$$|R_{n,n}(x)| \leq \sum_{k=1}^{n} A_k \left(\frac{C}{d}\right)^{2(n-k)} + \frac{a_1}{|\varphi(n)x - b_0|} \left(\frac{C}{d}\right)^{2n} .$$

Clearly $A_k \longrightarrow 0$ as k tends to infinity so that we can use the abelian theorem

$$a_n \to a \Longrightarrow \frac{1}{z^n} \sum_{k=1}^{n} a_k z^k \to \frac{z}{z-1} a \qquad (|z| > 1)$$

with $z = d/C$ and d large enough, to find

$$\lim_{n\to\infty} R_{n,n}(x) = 0 \qquad\qquad x \in [B+d,\infty) .$$

This means that for $x > B+d$

$$(4.3.10) \qquad \lim_{n \to \infty} \left\{ \frac{p_{n-1}(\varphi(n)x)}{p_n(\varphi(n)x)} - \frac{p_n(\varphi(n)x)}{p_{n+1}(\varphi(n)x)} \right\} = 0 .$$

Now consider the recurrence relation at $\varphi(n)x$ and divide everything by $p_n(\varphi(n)x)$, then we get

$$\frac{a_{n+1}}{\varphi(n)} \frac{p_{n+1}(\varphi(n)x)}{p_n(\varphi(n)x)} - \left(x - \frac{b_n}{\varphi(n)}\right) + \frac{a_n}{\varphi(n)} \frac{p_{n-1}(\varphi(n)x)}{p_n(\varphi(n)x)} = 0 .$$

Consider a subsequence of $p_{n-1}(\varphi(n)x)/p_n(\varphi(n)x)$ that converges to a limit $L(x)$, then by (4.3.1), (4.3.2) and (4.3.10) we find that L satisfies

$$\frac{a}{L} - (x - b) + aL = 0$$

from which we find

$$L(x) = \frac{x - b \pm \sqrt{(x-b)^2 - 4a^2}}{2a} .$$

Analyzing the behaviour for large x leads to the choice of the - sign so that every converging subsequence has the same limit

$$L(x) = \frac{x - b - \sqrt{(x-b)^2 - 4a^2}}{2a} = \frac{2a}{x - b + \sqrt{(x-b)^2 - 4a^2}}$$

for $x \in [B+d,\infty)$. We can extend this result to $x \in \mathbb{C}\setminus[A,B]$ by observing that for x in a compact set K in $\mathbb{C}\setminus[A,B]$ we have

$$\left|\frac{p_{n-1}(\varphi(n)x)}{p_n(\varphi(n)x)}\right| < \frac{1}{\varphi(n)} \sum_{j=1}^{n} \frac{a_{j,n}}{\left|x - \frac{x_{j,n}}{\varphi(n)}\right|} < \frac{a_n}{\varphi(n)} \frac{1}{\delta}$$

where δ is the distance from K to $[A,B]$, which is the set where all the contracted zeros $x_{j,n}/\varphi(n)$ are dense, so that $\{p_{n-1}(\varphi(n)x)/p_n(\varphi(n)x)\}$ is a sequence of analytic functions in $\mathbb{C}\setminus[A,B]$ that is uniformly bounded on compact subsets of $\mathbb{C}\setminus[A,B]$ and therefore one can use the Theorem of Stieltjes-Vitali. ∎

The rational function

$$\frac{\varphi(n)}{a_n}\,\frac{p_{n-1}(\varphi(n)x)}{p_n(\varphi(n)x)}$$

is the Stieltjes transform of the measure μ_n defined by

$$\begin{cases} \mu_n(\{\frac{x_{j,n}}{\varphi(n)}\}) = \lambda_{j,n}p^2_{n-1}(x_{j,n}) & j = 1,2,\ldots,n \\ \mu_n(A) = 0 & \text{if } A \text{ contains no zeros of } p_n(\varphi(n)x) \end{cases}$$

where $\{\lambda_{j,n} : j = 1,\ldots,n\}$ are the Christoffel numbers for the orthogonal polynomials $\{p_n(x)\}$. By Theorem 4.10 we therefore have for $a > 0$

$$\lim_{n\to\infty} S(\mu_n;x) = \frac{2}{x-b+\sqrt{(x-b)^2-4a^2}} = \frac{1}{2\pi a^2}\int_{b-2a}^{b+2a}\sqrt{4a^2-(t-b)^2}\,\frac{dt}{x-t}$$

uniformly on compact subsets of $\mathbb{C}\backslash[A,B]$, which by the theorem of Grommer and Hamburger (appendix) implies

$$\lim_{n\to\infty}\sum_{j=1}^{n}\lambda_{j,n}p^2_{n-1}(x_{j,n})\,f\left(\frac{x_{j,n}}{\varphi(n)}\right) = \frac{1}{2\pi a^2}\int_{b-2a}^{b+2a}\sqrt{4a^2-(t-b)^2}\,f(t)\,dt$$

for every bounded and continuous function f. Actually the result holds if f is Riemann integrable on $[b-2a,b+2a]$. If $a = 0$ then

$$\lim_{n\to\infty}\sum_{j=1}^{n}\lambda_{j,n}p^2_{n-1}(x_{j,n})\,f\left(\frac{x_{j,n}}{\varphi(n)}\right) = f(b)$$

for every bounded and measurable function that is continuous at b.

If one wants to study the asymptotic behaviour of the ratio

$$\frac{1}{n}\,\frac{\{p_n(\varphi(n)x)\}'}{p_n(\varphi(n)x)},$$

where the dash indicates a derivative with respect to x, then one needs to assume more than (4.3.1). The appropriate conditions seem to be in terms of *regularly varying functions*.

<u>Definition</u> : A non-negative function $\varphi : \mathbb{R}^+ \longrightarrow \mathbb{R}^+$ is regularly varying (at infinity) if for every $t > 0$

$$\lim_{x\to\infty}\frac{\varphi(xt)}{\varphi(x)} = h(t)$$

where h is a measurable function.

The limit function h obviously satisfies the functional equation $h(xy) = h(x)h(y)$ and the only measurable solution is $h(t) = t^\alpha$ for some $\alpha \in R$. So φ is regularly varying if

$$\lim_{x \to \infty} \frac{\varphi(xt)}{\varphi(x)} = t^\alpha \qquad\qquad t > 0$$

and α is called the exponent (or index) of regular variation. We refer to Seneta [169] and Bingham-Goldie-Teugels [28'] for more details about regular variation.

Theorem 4.11. (Van Assche [191]) : Suppose that (4.3.2) holds and that φ is regularly varying with exponent $\alpha > 0$. Then for $a > 0$

$$(4.3.11) \qquad \lim_{x \to \infty} \frac{1}{n} \frac{\{p_n(\varphi(n)x)\}'}{p_n(\varphi(n)x)} = \int_0^1 \frac{1}{\sqrt{(x-bt^\alpha)^2 - 4a^2t^{2\alpha}}} dt$$

uniformly on compact subsets of $\mathbb{C}\setminus[A,B]$, where $[A,B]$ is the smallest interval containing $\{0\}$ and $[b-2a,b+2a]$. If $a = 0$ then

$$(4.3.12) \qquad \lim_{n \to \infty} \frac{1}{n} \frac{\{p_n(\varphi(n)x)\}'}{p_n(\varphi(n)x)} = \int_0^1 \frac{dt}{x - bt^\alpha}$$

uniformly on compact subsets of $\mathbb{C}\setminus[A,B]$, where $[A,B] = [0,b]$ if $b > 0$ and $[A,B] = [b,0]$ if $b < 0$.

Proof : We will only consider the case $a > 0$. Obviously

$$\frac{1}{n} \frac{\{p_n(\varphi(n)x)\}'}{p_n(\varphi(n)x)} = \frac{1}{n} \sum_{k=1}^{n} \frac{\{p_k(\varphi(n)x)\}'}{p_k(\varphi(n)x)} - \frac{\{p_{k-1}(\varphi(n)x)\}'}{p_{k-1}(\varphi(n)x)}$$

$$= -\frac{1}{n} \sum_{k=1}^{n} \left\{\frac{p_{k-1}(\varphi(n)x)}{p_k(\varphi(n)x)}\right\}' \Big/ \left\{\frac{p_{k-1}(\varphi(n)x)}{p_k(\varphi(n)x)}\right\} .$$

We change this sum to an integral by setting $k-1 = [nt]$ where $[nt]$ denotes the integer part of nt. This gives

$$(4.3.13) \qquad \frac{1}{n} \frac{\{p_n(\varphi(n)x)\}'}{p_n(\varphi(n)x)} = -\int_0^1 \left\{\frac{p_{[nt]}(\varphi(n)x)}{p_{[nt]+1}(\varphi(n)x)}\right\}' \Big/ \left\{\frac{p_{[nt]}(\varphi(n)x)}{p_{[nt]+1}(\varphi(n)x)}\right\} dt .$$

In order to obtain the asymptotic behaviour of the ratio $p_{[nt]}(\varphi(n)x)/p_{[nt]+1}(\varphi(n)x)$ we need

$$\lim_{n\to\infty} \frac{a_{[nt]}}{\varphi(n)} = at^{\alpha} \quad , \quad \lim_{n\to\infty} \frac{b_{[nt]}}{\varphi(n)} = bt^{\alpha} \qquad t \in (0,1]$$

which follows from the regular variation of φ. Therefore by Theorem 4.10

$$\lim_{n\to\infty} \frac{p_{[nt]}(\varphi(n)x)}{p_{[nt]+1}(\varphi(n)x)} = \frac{x - bt^{\alpha} - \sqrt{(x - bt^{\alpha})^2 - 4a^2t^{2\alpha}}}{2at^{\alpha}} \qquad 0 < t \leq 1$$

uniformly on compact sets of $\mathbb{C}\setminus[A,B]$. One is allowed to take derivatives on both sides of this expression so that one also has

$$\lim_{n\to\infty} \left\{\frac{p_{[nt]}(\varphi(n)x)}{p_{[nt]+1}(\varphi(n)x)}\right\}' = \frac{x - bt^{\alpha} - \sqrt{(x-bt^{\alpha})^2 - 4a^2t^{2\alpha}}}{2at^{\alpha}} \frac{1}{\sqrt{(x-bt^{\alpha})^2 - 4a^2t^{2\alpha}}}$$

$$0 < t \leq 1$$

uniformly on compact subsets of $\mathbb{C}\setminus[A,B]$. One can show that Lebesgue's theorem of dominated convergence can be used in (4.3.13), giving (4.3.11).

■

The rational function

$$\frac{1}{n} \frac{\{p_n(\varphi(n)x)\}'}{p_n(\varphi(n)x)}$$

is the Stieltjes transform of the measure ν_n given by

$$\begin{cases} \nu_n(\{\frac{x_{j,n}}{\varphi(n)}\}) = \frac{1}{n} & j = 1,2,\dots,n \\ \nu_n(A) = 0 & A \text{ contains no zeros of } p_n(\varphi(n)x) \end{cases}$$

and by Theorem 4.11 we find that $S(\nu_n;x)$ converges uniformly on compact subsets of $\mathbb{C}\setminus[A,B]$ to some function $S(x)$. This limit has to be the Stieltjes transform of a probability measure by the theorem of Grommer and Hamburger (appendix). Actually the integrand in (4.3.11) is the Stieltjes transform of the arcsin measure on $[(b-2a)t^{\alpha},(b+2a)t^{\alpha}]$ so that

$$\int_0^1 \frac{dt}{\sqrt{(x-bt^\alpha)^2 - 4a^2t^{2\alpha}}} = \frac{1}{\pi} \int_0^1 \int_{(b-2a)t^\alpha}^{(b+2a)t^\alpha} \frac{1}{\sqrt{4a^2t^{2\alpha} - (y-bt^\alpha)^2}} \frac{dy}{x-y}\, dt .$$

We want to change the order of integration. In order to do this we have to distinguish between two cases. If $b^2 - 4a^2 < 0$ then $b - 2a < 0 < b + 2a$ and

$$\int_0^1 \frac{dt}{\sqrt{(x-bt^\alpha)^2 - 4a^2t^{2\alpha}}} = \int_{b-2a}^{b+2a} v(y) \frac{dy}{x-y}$$

where

$$v(y) = \begin{cases} \dfrac{1}{\pi} \displaystyle\int_{(\frac{y}{b+2a})^{1/\alpha}}^{1} \frac{dt}{\sqrt{4a^2t^{2\alpha} - (y-bt^\alpha)^2}} & 0 < y < b + 2a \\[2ex] \dfrac{1}{\pi} \displaystyle\int_{(\frac{y}{b-2a})^{1/\alpha}}^{1} \frac{dt}{\sqrt{4a^2t^{2\alpha} - (y-bt^\alpha)^2}} & b - 2a < y < 0 . \end{cases}$$

Notice that the weight function $v(y)$ corresponds to the weight function of the Ullman measure $\mu^{1/\alpha}$ if $a = 1/2$ and $b = 0$. If $b^2 - 4a^2 > 0$ then $b - 2a$ and $b + 2a$ have the same sign. Suppose that both quantities are positive (the other case can be handled similarly), then

$$\int_0^1 \frac{dt}{\sqrt{(x-bt^\alpha)^2 - 4a^2t^{2\alpha}}} = \int_0^{b+2a} v(y) \frac{dy}{x-y}$$

where

$$v(y) = \begin{cases} \dfrac{1}{\pi} \displaystyle\int_{(\frac{y}{b+2a})^{1/\alpha}}^{1} \frac{dt}{\sqrt{4a^2t^{2\alpha} - (y-bt^\alpha)^2}} & b - 2a < y < b + 2a \\[2ex] \dfrac{1}{\pi} \displaystyle\int_{(\frac{y}{b+2a})^{1/\alpha}}^{(\frac{y}{b-2a})^{1/\alpha}} \frac{dt}{\sqrt{4a^2t^{2\alpha} - (y-bt^\alpha)^2}} & 0 < y < b-2a , \end{cases}$$

As a consequence we have

$$\lim_{n\to\infty} \frac{1}{n} \sum_{j=1}^{n} f\left(\frac{x_{j,n}}{\varphi(n)}\right) = \int_A^B f(y)v(y)dy$$

$$= \frac{1}{\pi} \int_0^1 \int_{b-2a}^{b+2a} \frac{f(yt^\alpha)}{\sqrt{4a^2 - (y-b)^2}} \, dydt$$

for every bounded continuous function f. The measures with weight function v(y) can be considered as Ullman measures with parameter $1/\alpha$ for the interval [b-2a,b+2a]. Notice that the support of this measure is [b-2a,b+2a] if $b^2 - 4a^2 < 0$ but that the support is [0,b+2a] if $b - 2a > 0$ and [b-2a,0] if $b + 2a < 0$. If $a = 0$ then

$$\lim_{n \to \infty} \frac{1}{n} \sum_{j=1}^{n} f(\frac{x_{j,n}}{\varphi(n)}) = \int_0^1 f(bt^\alpha)dt$$

for every bounded and continuous function f. The special case $\alpha = 1$ means that the contracted zeros $\{x_{j,n}/\varphi(n)\}$ are uniformly distributed on [0,b] if $b > 0$ or on [b,0] if $b < 0$.

These results hold also for $\alpha = 0$ provided that φ is an increasing function. The Ullman measure μ^∞ on [-1,1] is equal to the arcsin measure on [-1,1] which explains why this arcsin measure appears for some contracted zeros (see e.g. Erdös [52], Erdös-Freud [53]).

4.4. Plancherel-Rotach asymptotics

Plancherel and Rotach [156] have studied the asymptotic behaviour of Hermite polynomials. They obtained the following results :

Theorem 4.12. (Plancherel-Rotach [156], Szegö [175] p. 201). Let ε and ω be fixed positive numbers, then for the Hermite polynomials $\{H_n(x) : n = 0,1,2,...\}$

$$e^{-(n+1/2)\cos^2\phi} H_n(\sqrt{2n+1}\cos\phi) = 2^{\frac{n}{2}+\frac{1}{4}} \sqrt{n!}\, (\pi n)^{-1/4} (\sin\phi)^{-1/2} \times \{\sin[(\frac{n}{2}+\frac{1}{4})(\sin 2\phi - 2\phi) + \frac{3\pi}{4}] + O(\frac{1}{n})\} \tag{4.4.1}$$

where the O-term holds uniformly for $\varepsilon < \phi < \pi-\varepsilon$, and

$$e^{-(n+1/2)\cosh^2\phi} H_n(\sqrt{2n+1}\cosh\phi) = 2^{\frac{n}{2}-\frac{3}{4}} \sqrt{n!}\, (\pi n)^{-1/4} (\sinh\phi)^{-1/2} \times \exp[(\frac{n}{2}+\frac{1}{4})(2\phi - \sinh 2\phi)]\{1 + O(\frac{1}{n})\} \tag{4.4.2}$$

where the O-term holds uniformly for $\varepsilon < \phi < \omega$.

Notice that in both results above the argument $\sqrt{2n+1}\, x$ of the Hermite polynomial H_n changes as the degree of the polynomial changes, so that the zeros of the polynomial $H_n(\sqrt{2n+1}\, x)$ are exactly the contracted zeros $x_{j,n}/\sqrt{2n+1}$, with $x_{j,n}$ the zeros of the Hermite polynomial H_n. Asymptotic results for a sequence $\{p_n(c_n x) : n = 0,1,2,\ldots\}$ where c_n is of the same order as the largest zero $x_{n,n}$ of p_n, are nowadays known as *Plancherel-Rotach asymptotics*. Plancherel and Rotach found their result by using the generating function for the Hermite polynomials. Another possible proof uses the differential equation that is satisfied by the Hermite polynomials. There are always two different cases when one analyzes the asymptotic behaviour of orthogonal polynomials, in particular for Plancherel-Rotach asymptotics. Suppose that one wants to find the asymptotic behaviour of the sequence $\{p_n(c_n x) : n = 1,2,\ldots\}$, then this behaviour is oscillatory when x is in the set [A,B] where the contracted zeros are dense (see (4.4.1)) and the behaviour is exponential on $\mathbb{C}\setminus[A,B]$ (see (4.4.2)).

When one considers the weight functions e^{-x^4} and e^{-x^6} (or weight functions that look like these) then it is possible to find a second order differential equation for the orthogonal polynomials which can then be analyzed. This approach has been taken by Nevai, Sheen and Bouldry to find Plancherel-Rotach asymptotics in the oscillatory region for these orthogonal polynomials :

Theorem 4.13. (Nevai [145],[147]). Let $\{p_n(x) : n = 1,2,\ldots\}$ be the orthogonal polynomials corresponding to the weight function e^{-x^4} and let $0 < \varepsilon < \pi/2$ be fixed, then

$$(4.4.3)\qquad e^{-(\frac{2n}{3})\cos^4\phi}\, p_n\!\left((\tfrac{4n}{3})^{1/4}\cos\phi\right) = 12^{1/8}\,\pi^{-1/2}\,n^{-1/8}\,(\sin\phi)^{-1/2}$$

$$\times\ \{\cos[\tfrac{n}{12}\,(12\phi - 4\sin 2\phi - \sin 4\phi) + \tfrac{\phi}{2} - \tfrac{\pi}{4}] + O(\tfrac{1}{n})\}$$

holds uniformly for $\varepsilon < \phi < \pi-\varepsilon$.

Theorem 4.14. (Sheen, see [148],[149],[151]). Let $\{p_n(x) : n = 1,2,\ldots\}$ be the orthogonal polynomials corresponding to $e^{-x^6/6}$ and let $0 < \varepsilon < \frac{\pi}{2}$ be fixed, then

$$(4.4.4)\qquad e^{-(\frac{8n}{15})\cos^6\phi}\, p_n\!\left((\tfrac{32n}{5})^{1/6}\cos\phi\right) = 10^{1/12}\,\pi^{-1/2}\,n^{-1/12}\,(\sin\phi)^{-1/2}$$

$$\times\left\{\cos\,[\tfrac{n}{60}\,(60\phi - 15\sin 2\phi - 6\sin 4\phi - \sin 6\phi) + \tfrac{\phi}{2} - \tfrac{\pi}{4}] + O(\tfrac{1}{n})\right\}$$

holds uniformly for $\varepsilon < \phi < \pi-\varepsilon$.

Bauldry [19] used the same techniques, but needed much more elaborate calculations, to obtain Plancherel-Rotach asymptotics in the oscillatory region for orthogonal polynomials corresponding to the weight function $e^{-\pi_4(x)}$, where $\pi_4(x) = \frac{x^4}{3} + q_3\frac{x^3}{3} + q_2\frac{x^2}{2} + q_1 x$ is a polynomial of degree 4.

Lubinsky and Saff [110] were able to find asymptotic results for a very general class of weights. Alternatively one can find Plancherel-Rotach asymptotics in $\mathbb{C}\setminus[A,B]$ when one uses the recurrence relation

$$xp_n(x) = a_{n+1}p_{n+1}(x) + b_n p_n(x) + a_n p_{n-1}(x) \qquad n = 0,1,2,\ldots$$

rather than the weight function. The proper conditions again seem to be in terms of regular variation :

Theorem 4.15. (Van Assche - Geronimo [195]). Suppose that $\{c_n : n = 1,2,\ldots\}$ is a regularly varying sequence with index $\alpha > 0$ such that

$$(4.4.5) \qquad \lim_{n\to\infty} \frac{a_n}{c_n} = a > 0 \quad , \quad \lim_{n\to\infty} \frac{b_n}{c_n} = b \in \mathbb{R}$$

and

$$(4.4.6) \qquad \lim_{n\to\infty} n\left(\frac{a_{n+1} - a_n}{c_n}\right) = a\alpha \quad , \quad \lim_{n\to\infty} n\left(\frac{b_{n+1} - b_n}{c_n}\right) = b\alpha \ .$$

If $z_{k,n} = \rho\left(\frac{c_n x - b_k}{2a_k}\right)$ and $\rho(x) = x + \sqrt{x^2 - 1}$, then

$$(4.4.7) \qquad \lim_{n\to\infty} \frac{p_n(c_n x)}{\prod_{k=1}^{n} z_{k,n}} = \left\{\frac{(x-b)^2 - 4a^2}{x^2}\right\}^{-1/4} \exp\left\{\frac{b}{2}\int_0^1 \frac{ds}{\sqrt{(x-bs)^2 - 4a^2s^2}}\right\}$$

uniformly on compact subsets of $\mathbb{C}\setminus[A,B]$, where $[A,B]$ is the smallest interval that contains $\{0\}$ and $[b-2a,b+2a]$.

Proof : (sketch) as in Theorem 2.28 one introduces

$$\phi_{k,n}(x) = p_k(c_n x) - \frac{1}{z_{k,n}} p_{k-1}(c_n x)$$

then one easily shows that

$$\phi_{k+1,n}(x) - \frac{a_k}{a_{k+1}} z_{k,n}\phi_{k,n}(x) = \Delta_{k,n} p_k(c_n x)$$

where

$$\Delta_{k,n} = \frac{a_k}{a_{k+1}} \frac{1}{z_{k,n}} - \frac{1}{z_{k+1,n}} .$$

From this one finds

$$(4.4.8)\qquad \frac{a_{n+1}}{a_1} \frac{\phi_{n+1,n}(x)}{\phi_{1,n}(x) \prod_{k=1}^{n} z_{k,n}} = \prod_{k=1}^{n} \left\{ 1 + \frac{\Delta_{k,n}}{\frac{a_k}{a_{k+1}} \left(z_{k,n} - \frac{p_{k-1}(c_n x)}{p_k(c_n x)} \right)} \right\}$$

$$= \exp \left\{ n \int_0^1 \log \left\{ 1 + \frac{\Delta_{[nt]+1,n}}{\frac{a_{[nt]+1}}{a_{[nt]+2}} \left(z_{[nt]+1,n} - \frac{p_{[nt]}(c_n x)}{p_{[nt]+1}(c_n x)} \right)} \right\} dt \right\} .$$

By (4.4.5) and (4.4.6) one has

$$\lim_{n \to \infty} n \Delta_{[nt]+1,n} = \frac{\alpha}{2at} \left\{ b - \frac{b(x - bt^\alpha) + 4a^2 t^\alpha}{\sqrt{(x - bt^\alpha)^2 - 4a^2 t^{2\alpha}}} \right\}$$

$$\lim_{n \to \infty} z_{[nt]+1,n} = \frac{x - bt^\alpha}{2at^\alpha} + \sqrt{\left(\frac{x - bt^\alpha}{2at^\alpha}\right)^2 - 1}$$

and by Theorem 4.10

$$\lim_{n \to \infty} \frac{p_{[nt]}(c_n x)}{p_{[nt]+1}(c_n x)} = \frac{x - bt^\alpha}{2at^\alpha} - \sqrt{\left(\frac{x - bt^\alpha}{2at^\alpha}\right)^2 - 1}$$

uniformly for x on compact sets of $\mathbb{C}\backslash[A,B]$. Interchanging limit and integral on the right hand side of (4.4.8) gives

$$\lim_{n \to \infty} \frac{\phi_{n+1,n}(x)}{\prod_{k=1}^{n} z_{k,n}} = \frac{x}{a} \left\{ \frac{(x-b)^2 - 4a^2}{x^2} \right\}^{1/4} \exp \left\{ \frac{b}{2} \int_0^1 \frac{ds}{\sqrt{(x - bs)^2 - 4a^2 s^2}} \right\} .$$

Observe that

$$\frac{\phi_{n+1,n}(x)}{p_n(c_n x)} = \frac{p_{n+1}(c_n x)}{p_n(c_n x)} - \frac{1}{z_{n+1,n}}$$

so that

$$\lim_{n \to \infty} \frac{\phi_{n+1,n}(x)}{p_n(c_n x)} = 2\sqrt{\left(\frac{x-b}{2a}\right)^2 - 1}$$

which gives (4.4.7). ∎

If $a_n = an^\alpha$ and $b_n = bn^\alpha$ then one can analyze the product $\prod_{k=1}^{n} z_{k,n}$ in more detail and the result is

$$(4.4.9) \qquad \lim_{n \to \infty} (2\pi n)^{\alpha/2} \frac{p_n(n^\alpha x)}{z^n H^n} = (az)^{1/2} \{(x-b)^2 - 4a^2\}^{-1/4} \exp\left\{\frac{b}{2}\int_0^1 \frac{ds}{\sqrt{(x-bs)^2 - 4a^2 s^2}}\right\}$$

uniformly on compact sets of $\mathbb{C}\setminus[A,B]$, where

$$z = \frac{x-b}{2a} + \sqrt{\left(\frac{x-b}{2a}\right)^2 - 1}$$

$$H = \exp\left\{\alpha x \int_0^1 \frac{dt}{\sqrt{(x - bt^\alpha)^2 - 4a^2 t^{2\alpha}}}\right\}.$$

Note that the special case $\alpha = 1/2$ and $b = 0$ in (4.4.9) corresponds exactly to the result for Hermite polynomials in Theorem 3.5.

4.5. Weighted zero distributions

In this section we will investigate the zeros $\{x_{j,n} : j = 1,2,\ldots,n;\ n = 1,2,\ldots\}$ of orthogonal polynomials $\{p_n(x) : n = 1,2,\ldots\}$ by means of a sequence of measures $\{\xi_n : n = 1,2,\ldots\}$ for which

$$\begin{cases} \xi_n(\{x_{j,n}\}) = \beta_{j,n} > 0 & j = 1,2,\ldots,n \\ \xi_n(A) = 0 & A \text{ contains no zeros of } p_n. \end{cases}$$

We will choose the weights $\{\beta_{j,n} : j = 1,2,\ldots,n\}$ in such a way that large zeros (in absolute value) receive a small weight. This method was proposed in [187]. We start with a weight function w that satisfies (4.1.10) and use a result of Rakhmanov :

Theorem 4.16. (Rakhmanov [163]) : For orthogonal polynomials $\{p_n(x) : n = 0,1,2,...\}$ with a weight function w that is almost everywhere positive on $(-\infty,\infty)$ and that satisfies (4.1.10) with $\alpha > 1$ we have, uniformly on compact sets of $\mathbb{C}\backslash\mathbb{R}$

$$(4.5.1) \qquad \lim_{n\to\infty} \frac{\log|p_n(z)|}{n^{1-1/\alpha}} = \frac{\alpha}{\alpha-1}\left(\frac{\lambda_\alpha}{2}\right)^{1/\alpha}|\mathrm{Im}\, z|$$

where λ_α is given by (4.1.4).

Let $\alpha > 1$ and define a sequence of *weighted zero distributions* $\{\xi_n^\alpha : n = 1,2,...\}$ by

$$\begin{cases} \xi_n^\alpha(\{x_{j,n}\}) = \dfrac{\alpha-1}{\alpha}\left(\dfrac{2}{\lambda_\alpha}\right)^{1/\alpha} \dfrac{n^{-1+1/\alpha}}{1+x_{j,n}^2} & j = 1,2,...,n \\ \xi_n^\alpha(A) = 0 & \text{if } A \text{ contains no zeros of } p_n \end{cases}$$

then we have

Theorem 4.17. (Van Assche [187]) : If w is a weight function that is positive almost everywhere on $(-\infty,\infty)$ and for which (4.1.10) is valid with $\alpha > 1$, then ξ_n^α converges weakly to the *Cauchy measure* ξ for which

$$(4.5.3) \qquad \xi(A) = \frac{1}{\pi}\int_A \frac{1}{1+x^2}\,dx$$

for every Borel set A in $\mathbb{R}$.

Proof : We define another sequence of measures by

$$\begin{cases} \nu_n^\alpha(\{x_{j,n}\}) = \dfrac{1}{n^{1-1/\alpha}} & j = 1,2,...,n \\ \nu_n^\alpha(A) = 0 & A \text{ contains no zeros of } p_n \end{cases}$$

then (4.5.1) is the same as

$$(4.5.4) \qquad \lim_{n\to\infty}\left\{\frac{\log k_n}{n^{1-1/\alpha}} + \int_{-\infty}^{\infty} \log|z-x|\,d\nu_n^\alpha(x)\right\} = \frac{\alpha}{\alpha-1}\left(\frac{\lambda_\alpha}{2}\right)^{1/\alpha}|\mathrm{Im}\, z|$$

uniformly on compact subsets of $\mathbb{C}\setminus\mathbb{R}$. Both sides of this asymptotic relation are harmonic functions in the sets $\{\operatorname{Im} z > 0\}$ and $\{\operatorname{Im} z < 0\}$. If we let $z = u + iv$ ($u,v \in \mathbb{R}$) then the partial derivatives $\frac{\partial}{\partial u}$ and $\frac{\partial}{\partial v}$ of the left hand side of (4.5.4) will converge to the partial derivatives of the right hand side, uniformly on compact sets of $\mathbb{C}\setminus\mathbb{R}$ (Kellogg [101], p. 249), so that

$$\lim_{n\to\infty} \int_{-\infty}^{\infty} \frac{u-x}{(u-x)^2+v^2}\, d\nu_n^\alpha(x) = 0$$

$$\lim_{n\to\infty} \int_{-\infty}^{\infty} \frac{v}{(u-x)^2+v^2}\, d\nu_n^\alpha(x) = \frac{\alpha}{\alpha-1} \left(\frac{\lambda\alpha}{2}\right)^{1/\alpha} \operatorname{sgn} v ,$$

where $\operatorname{sgn} v = 1$ if $v > 0$ and $\operatorname{sgn} v = -1$ if $v < 0$ and the convergence is uniform on compact sets of $\mathbb{C}\setminus\mathbb{R}$. Hence

$$(4.5.5) \qquad \int_{-\infty}^{\infty} \frac{1}{z-x}\, d\nu_n^\alpha(x) = \int_{-\infty}^{\infty} \frac{u-x-iv}{(u-x)^2+v^2}\, d\nu_n^\alpha(x) \longrightarrow -i\, \frac{\alpha}{\alpha-1} \left(\frac{\lambda\alpha}{2}\right)^{1/\alpha} \operatorname{sgn} v .$$

and in particular

$$(4.5.6) \qquad \lim_{n\to\infty} \int_{-\infty}^{\infty} \frac{1}{i \pm x}\, d\nu_n^\alpha(x) = \mp i\, \frac{\alpha}{\alpha-1} \left(\frac{\lambda\alpha}{2}\right)^{1/\alpha}$$

from which we find

$$\xi_n^\alpha(\mathbb{R}) = \frac{\alpha-1}{\alpha} \left(\frac{2}{\lambda\alpha}\right)^{1/\alpha} \int_{-\infty}^{\infty} \frac{1}{1+x^2}\, d\nu_n^\alpha(x) \longrightarrow 1 .$$

Next we will calculate the Stieltjes transform of ξ_n^α :

$$S(\xi_n^\alpha;z) = \frac{\alpha-1}{\alpha} \left(\frac{2}{\lambda\alpha}\right)^{1/\alpha} \int_{-\infty}^{\infty} \frac{1}{1+x^2}\, \frac{d\nu_n^\alpha(x)}{z-x}$$

$$= \frac{\alpha-1}{\alpha} \left(\frac{2}{\lambda\alpha}\right)^{1/\alpha} \frac{1}{1+z^2} \left\{ \int_{-\infty}^{\infty} \frac{d\nu_n^\alpha(x)}{z-x} + z \int_{-\infty}^{\infty} \frac{d\nu_n^\alpha(x)}{1+x^2} + \int_{-\infty}^{\infty} \frac{x}{1+x^2}\, d\nu_n^\alpha(x) \right\} .$$

Using (4.5.5) and (4.5.6) gives

$$\lim_{n\to\infty} S(\xi_n^\alpha;z) = \frac{1}{1+z^2}(z - i \operatorname{sgn} v) = \begin{cases} \dfrac{1}{z+i} & \text{if } \operatorname{Im} z > 0 \\ \\ \dfrac{1}{z-i} & \text{if } \operatorname{Im} z < 0 . \end{cases}$$

The limit is the Stieltjes transform of the Cauchy measure given in (4.5.3) (see appendix) so that the result follows from the theorem of Grommer and Hamburger. ∎

As a consequence we have that for every bounded and continuous function f

$$\lim_{n\to\infty} \frac{\alpha-1}{\alpha}\left(\frac{2}{\lambda\alpha}\right)^{1/\alpha} \frac{1}{n^{1-1/\alpha}} \sum_{j=1}^{n} \frac{f(x_{j,n})}{1+x_{j,n}^2} = \frac{1}{\pi}\int_{-\infty}^{\infty} \frac{f(x)}{1+x^2}\,dx \tag{4.5.7}$$

and this holds even if f is a bounded measurable function with discontinuities on a set of Lebesgue measure zero.

If $\{p_n^+(x) : n = 1,2,\ldots\}$ is a sequence of orthogonal polynomials on $[0,\infty)$ with zeros $\{x_{j,n}^+ : j = 1,2,\ldots,n;\ n = 1,2,\ldots\}$ then we introduce another sequence of weighted zero distributions

$$\begin{cases} \xi_n^{\gamma+}(\{x_{j,n}^+\}) = \dfrac{2\gamma-1}{\gamma}\left(\dfrac{2}{\lambda_{2\gamma}}\right)^{1/2\gamma} \dfrac{(2n)^{-1+1/2\gamma}}{1+x_{j,n}^+} & j = 1,2,\ldots,n \\ \\ \xi_n^{\gamma+}(A) = 0 & \text{if } A \text{ contains no zeros of } p_n^+ . \end{cases}$$

A result, analoguous with Theorem 4.17 is

Theorem 4.18. (Van Assche [187]) : If w^+ is a weight function on $[0,\infty)$ that is positive almost everywhere on $[0,\infty)$ and for which

$$\lim_{n\to\infty} x^{-\gamma} \log w^+(x) = -1 \qquad \gamma > \frac{1}{2}$$

then $\xi_n^{\gamma+}$ converges weakly to the measure ξ^+ for which

$$\xi^+(A) = \frac{1}{\pi}\int_A \frac{dt}{\sqrt{t}(1+t)} \tag{4.5.9}$$

where A is a Borel set in $[0,\infty)$.

Proof : Consider the weight function $w(x) = |x|w^{+}(x^{2})$ on $(-\infty,\infty)$. This weight function satisfies the conditions of Theorem 4.17 with $\alpha = 2\gamma$. The zeros $\{x_{j,n}\}$ of the orthogonal polynomials corresponding with the weight function w are related to the zeros $\{x_{j,n}^{+}\}$ by

$$x_{j,n}^{+} = x_{j,2n}^{2} \qquad j = 1,2,\ldots,n$$

so that for $x > 0$

$$\xi_{n}^{\gamma+}((0,x]) = 2\xi_{2n}^{2\gamma}((0,\sqrt{x}])$$

If we let n tend to infinity, then

$$\xi^{+}((0,x]) = \frac{2}{\pi}\int_{0}^{\sqrt{x}} \frac{dt}{1+t^{2}} = \frac{1}{\pi}\int_{0}^{x} \frac{dy}{\sqrt{y}\,(1+y)}$$

which gives the result. ∎

If f is a bounded continuous function on $[0,\infty)$ then it follows that

$$\lim_{n\to\infty} \frac{2\gamma-1}{\gamma}\left(\frac{2}{\lambda_{2\gamma}}\right)^{1/2\gamma} \frac{1}{(2n)^{1-1/2\gamma}} \sum_{j=1}^{n} \frac{f(x_{j,n}^{+})}{1+x_{j,n}^{+}} = \frac{1}{\pi}\int_{0}^{\infty} \frac{f(t)}{\sqrt{t}(1+t)}\,dt$$

and again we may allow f to be a bounded measurable function with discontinuities on a set of Lebesgue measure zero.

4.6. Rate of convergence for weighted zero distributions

The weight function (0.1.7) for generalized Laguerre polynomials satisfies the conditions of Theorem 4.18 (with $\gamma = 1$) so that the sequence $\{\xi_{n}^{1+} : n = 1,2,\ldots\}$ converges weakly to ξ^{+}. In order to find the rate of convergence we need an asymptotic formula which is stronger than Theorem 4.16. A useful result is *Perron's formula* :

Theorem 4.19. (Szegö [175], Thm. 8.22.3) : For every positive integer p and for $\alpha > -1$ we have

$$L_n^\alpha(z) = \frac{1}{2\sqrt{\pi}} e^{z/2}(-z)^{-(2\alpha+1)/4} n^{(2\alpha-1)/4} \exp(2\sqrt{-nz}) \tag{4.6.1}$$

$$\times \left\{ \sum_{j=0}^{p-1} C_j(\alpha;z)n^{-j/2} + O(n^{-p/2}) \right\}$$

where the O-term is uniformly bounded on compact subsets of $\mathbb{C}\setminus[0,\infty)$ and the roots $(-z)^{-(2\alpha+1)/4}$ and $\sqrt{-z}$ are real and positive for $z < 0$.

In this formula $C_0(\alpha;z) = 1$ but in order to obtain the rate of convergence we also need $C_1(\alpha;z)$. A careful analysis of the steepest descent method in Szegö [175] results in

$$C_1(\alpha;z) = \frac{1}{4\sqrt{-z}} \left(-3z + \frac{1}{3} z^2 + \frac{1}{4} - \alpha^2\right) \tag{4.6.2}$$

(details are worked out in [187]). In this section we will use two square roots : let $z = re^{i\theta}$, then

$$\begin{cases} z^{1/2} = \sqrt{r}\, e^{i\theta/2} & \theta \in [0,2\pi) \\ \sqrt{z} = \sqrt{r}\, e^{i\theta/2} & \theta \in [-\pi,\pi) \end{cases} \tag{4.6.3}$$

Notice that one always has $\sqrt{-z} = -iz^{1/2}$. The following theorem gives rate of convergence for weighted zero distributions involving the zeros of generalized Laguerre polynomials, in terms of Stieltjes transforms.

Theorem 4.20. (Van Assche [187]) : Let $\{\xi_n^+ : n = 1,2,\ldots\}$ and ξ^+ be the weighted zero distributions in (4.5.8) and (4.5.9) for zeros of generalized Laguerre polynomials $\{L_n^{(\alpha)}(x) : n = 0,1,2,\ldots\}$ ($\gamma = 1$, $\alpha > -1$), then

$$\lim_{n\to\infty} S(\sqrt{n}(\xi_n^+ - \xi^+);z) = -\frac{2\alpha+1}{4}\frac{1}{z} \tag{4.6.4}$$

uniformly on compact subsets of $\mathbb{C}\setminus[0,\infty)$.

Proof : Let $p_n(x) = L_n^{(\alpha)}(x)$. By means of the identity

$$\frac{d}{dz} L_n^{(\alpha)}(z) = -L_{n+1}^{(\alpha+1)}(z) \tag{4.6.5}$$

([175], p. 102) and Perron's formula (4.6.1) we have for $z \in \mathbb{C}\setminus[0,\infty)$

$$\frac{1}{\sqrt{n}}\frac{p_n'(z)}{p_n(z)} = -\frac{1}{\sqrt{-z}}\left(1-\frac{1}{n}\right)^{(2\alpha+1)/4} \exp\left\{2\sqrt{-nz}\left[\left(1-\frac{1}{n}\right)^{1/2}-1\right]\right\}$$

$$\times \left\{1+\frac{1}{\sqrt{n-1}}\,C_1(\alpha+1;z)\right\}\left\{1-\frac{1}{\sqrt{n}}\,C_1(\alpha;z)\right\} + o\left(\frac{1}{\sqrt{n}}\right)$$

Since

$$\left(1-\frac{1}{n}\right)^{(2\alpha+1)/4} = 1-\frac{2\alpha+1}{4n}+o\left(\frac{1}{n}\right)$$

$$\exp\left\{2\sqrt{-nz}\left[\left(1-\frac{1}{n}\right)^{1/2}-1\right]\right\} = 1-\frac{\sqrt{-z}}{\sqrt{n}}+o\left(\frac{1}{\sqrt{n}}\right)$$

this leads to

$$\text{(4.6.6)}\qquad \frac{1}{\sqrt{n}}\frac{p_n'(z)}{p_n(z)} = -\frac{1}{\sqrt{-z}}\left\{1+\frac{1}{\sqrt{n}}\left[C_1(\alpha+1;z)-C_1(\alpha;z)-\sqrt{-z}\right]\right\}+o\left(\frac{1}{\sqrt{n}}\right).$$

The Stieltjes transform of ξ_n^+ is given by

$$S(\xi_n^+;z) = \frac{1}{1+z}\left\{\xi_n^+(\mathbb{R}) + \int_0^\infty \frac{1+t}{z-t}\,d\xi_n^+(t)\right\}$$

but on the other hand we also have

$$\frac{p_n'(z)}{p_n(z)} = \sum_{j=1}^{n}\frac{1}{z-x_{j,n}^+} = \sqrt{n}\int_0^\infty \frac{1+t}{z-t}\,d\xi_n^+(t)\ .$$

Putting $z = -1$ gives

$$\xi_n^+(\mathbb{R}) = -\frac{1}{\sqrt{n}}\frac{p_n'(-1)}{p_n(-1)}\ .$$

The Stieltjes transform of ξ^+ is

$$S(\xi^+;z) = \frac{1}{1+z}-\frac{1}{\sqrt{-z}\,(1+z)} \qquad z \in \mathbb{C}\setminus[0,\infty)$$

(see appendix) and a combination of these formulas gives

$$S(\sqrt{n}(\xi_n^+ - \xi^+);z) = \frac{n}{1+z}\left\{\frac{1}{\sqrt{n}}\frac{p_n'(z)}{p_n(z)} + \frac{1}{\sqrt{-z}} - \frac{1}{\sqrt{n}}\frac{p_n'(-1)}{p_n(-1)} - 1\right\} .$$

If we use (4.6.6) then this becomes

$$S(\sqrt{n}(\xi_n^+ - \xi^+);z) = \frac{1}{1+z}\left\{C_1(\alpha+1;-1) - C_1(\alpha;-1) - \frac{1}{\sqrt{-z}}C_1(\alpha+1;z) + \frac{1}{\sqrt{-z}}C_1(\alpha;z)\right\} + o(1) .$$

The explicit expression (4.6.2) for C_1 then gives the desired result.

The weight function (0.1.12) for generalized Hermite polynomials $\{H_n^{(\beta)}(x) : n = 0,1,2,\ldots\}$ satisfies the conditions of Theorem 4.17 (with $\alpha = 2$). The rate of convergence of the weighted zero distributions ξ_n for the zeros of these Hermite polynomials to the Cauchy measure ξ in terms of Stieltjes transforms can be found in a similar way as in Thm. 4.20 and is given by

Theorem 4.21. (Van Assche [187]) : Let $\{\xi_n : n = 1,2,\ldots\}$ be the weighted zero distribution defined in (4.6.2) (with $\alpha = 2$) for the zeros of the Hermite polynomials $\{H_n^{(\beta)}(x) : n = 1,2,\ldots\}$ $(\beta > -1/2)$ and let ξ be the Cauchy measure (4.6.3), then

(4.6.7) $$\lim_{n\to\infty} S(\sqrt{2n}(\xi_n - \xi);z) = -\frac{\beta}{z}$$

uniformly on compact subsets of $\mathbb{C}\setminus\mathbb{R}$.

One has to be very careful and not conclude from Theorem 4.20 and 4.21 that the signed measures $K_n^+ = \sqrt{n}(\xi_n^+ - \xi^+)$ and $K_n = \sqrt{2n}(\xi_n - \xi)$ converge weakly. This is not true because the total variation of the measures K_n^+ and K_n are not uniformly bounded so that the theorem of Grommer and Hamburger (appendix) does not apply. A weaker result that looks very much like weak convergence is

Theorem 4.22. (Van Assche [187]) :

(i) Let ξ_n^+ and ξ^+ be the weighted zero distributions for the zeros of generalized Laguerre polynomials $\{L_n^{(\alpha)}(x) : n = 0,1,2,\ldots\}$ and let $K_n^+ = \sqrt{n}(\xi_n^+ - \xi^+)$.

If f is a function such that $f(\frac{1+y}{1-y})$ is analytic in an open set that contains [-1,1], then

(4.6.8) $$\lim_{n\to\infty}\left\{\sum_{j=1}^{n}\frac{f(x_{j,n}^{+})}{1+x_{j,n}^{+}}-\frac{\sqrt{n}}{\pi}\int_0^\infty \frac{f(x)}{\sqrt{x}\,(1+x)}\,dx\right\}$$

$$=\lim_{n\to\infty}\int_0^\infty f(x)dK_n^{+}(x) = -\frac{2\alpha+1}{4}f(0)-f(\infty)\ .$$

(ii) Let ξ_n and ξ be the weighted zero distributions for the zeros of generalized Hermite polynomials $\{H_n^{(\beta)}(x) : n = 0,1,2,\ldots\}$ and let $K_n = \sqrt{2n}(\xi_n - \xi)$. If f is a function such that $f(\pm\{\frac{1+y}{1-y}\}^{1/2})$ is analytic in an open set that contains [-1,1], then

$$\lim_{n\to\infty}\left\{\sum_{j=1}^{n}\frac{f(x_{j,n})}{1+x_{j,n}^{2}}-\frac{\sqrt{2n}}{\pi}\int_{-\infty}^\infty \frac{f(x)}{1+x^2}\,dx\right\}$$

$$=\lim_{n\to\infty}\int_{-\infty}^\infty f(x)dK_n(x) = -\beta f(0)-f(\infty)-f(-\infty)\ .$$

Proof : We will only consider (i), a proof of (ii) can be given along the same lines. Let $g(y) = f(\frac{1+y}{1-y})$, then g will be analytic in an open set that contains [-1,1]. Let Γ be a closed contour in that open set that encircles the interval [-1,1] then by Cauchy's theorem

$$\int_0^\infty f(x)dK_n^{+}(x) = \int_{-1}^{1} g(y)dK_n^{+}(\tfrac{1+y}{1-y})$$

$$=\frac{1}{2\pi i}\int_\Gamma g(z)\int_{-1}^{1}\frac{dK_n^{+}(\frac{1+y}{1-y})}{z-y}\,dz\ .$$

By a simple substitution this gives

$$\int_0^\infty f(x)dK_n^{+}(x) = \frac{1}{2\pi i}\int_\Gamma \frac{g(z)}{z-1}\int_0^\infty \frac{(1+x)dK_n^{+}(x)}{x-\frac{1+z}{1-z}}\,dz$$

$$=\frac{2}{2\pi i}\int_\Gamma \frac{g(z)}{(z-1)^2}S(K_n^{+};\tfrac{1+z}{1-z})dz + K_n^{+}(\mathbb{R})\frac{1}{2\pi i}\int_\Gamma \frac{g(z)}{z-1}\,dz\ .$$

Use Theorem 4.20 (which is possible since Γ is compact), then

$$\lim_{n\to\infty}\int_0^\infty f(x)dK_n^+(x) = -\frac{2\alpha+1}{2}\frac{1}{2\pi i}\int_\Gamma \frac{g(z)}{1-z^2}\,dz$$

$$+\left(\lim_{n\to\infty} K_n^+(\mathbb{R})\right)\frac{1}{2\pi i}\int_\Gamma \frac{g(z)}{z-1}\,dz\ .$$

By the theorem of residues and

$$K_n^+(\mathbb{R}) = \sqrt{n}\left\{\xi_n^+(\mathbb{R}) - \xi^+(\mathbb{R})\right\}$$

$$= -\sqrt{n}\left\{\frac{1}{\sqrt{n}}\frac{p_n'(-1)}{p_n(-1)} + 1\right\}$$

$$= C_1(\alpha+1;-1) - C_1(\alpha;-1) + o(1)$$

$$\longrightarrow -\frac{2\alpha-5}{4}$$

we find

$$\lim_{n\to\infty}\int_0^\infty f(x)dK_n^+(x) = -\frac{2\alpha+1}{4}\{g(-1) - g(1)\} - \frac{2\alpha+5}{4}g(1)$$

which gives (4.6.8) if one takes into account that $g(1) = f(\infty)$ and $g(-1) = f(0)$.

∎

CHAPTER 5 : ZERO DISTRIBUTION AND CONSEQUENCES

As in the previous chapters we define a sequence of discrete measures $\{\nu_n : n = 1,2,\ldots\}$ that describe the distribution of the zeros $\{x_{j,n} : j = 1,\ldots,n\}$ of the polynomials $\{p_n(x) : n = 1,2,3,\ldots\}$:

$$\begin{cases} \nu_n(\{\frac{x_{j,n}}{c_n}\}) = \frac{1}{n} & j = 1,2,\ldots,n \\ \nu_n(A) = 0 & A \text{ contains no zeros of } p_n(c_n x) . \end{cases}$$

This includes contracted zero distributions if one considers increasing sequences $\{c_n : n = 1,2,\ldots\}$. If the measure ν_n converges weakly to some probability measure ν then ν gives the *asymptotic distribution* of the zeros $\{x_{j,n} : j = 1,\ldots,n;\ n = 1,2,\ldots\}$. The existence and precise nature of this asymptotic distribution has been discussed in the previous chapters. In this chapter we give some consequences of the existence of the asymptotic zero distribution.

5.1. Asymptotic behaviour of the Taylor coefficients

Let $\{\hat{p}_n(x) : n = 1,2,\ldots\}$ be a sequence of monic polynomials (not necessarily orthogonal) then we denote the Taylor series of $\hat{p}_n(x)$ around $x = 0$ by

$$(5.1.1) \qquad \hat{p}_n(x) = \sum_{j=0}^{n} a_{j,n} x^{n-j} \qquad (a_{0,n} = 1) .$$

It is well known that the Taylor coefficients $\{a_{j,n} : j = 1,2,\ldots,n\}$ are related to the zeros $\{x_{j,n} : j = 1,\ldots,n\}$ by the *elementary symmetric relations of Viète*

$$a_{j,n} = \frac{(-1)^j}{j!} \sum_{\alpha_1 \neq \alpha_2 \neq \ldots \neq \alpha_j} x_{\alpha_1,n}\, x_{\alpha_2,n} \cdots x_{\alpha_j,n} .$$

If we suppose that the zeros of $\hat{p}_n$ are negative then this becomes

$$a_{j,n} = \frac{1}{j!} \sum_{\alpha_1 \neq \alpha_2 \neq \ldots \neq \alpha_j} |x_{\alpha_1,n} x_{\alpha_2,n} \ldots x_{\alpha_j,n}|$$

(5.1.2)

$$= \sum_{\alpha_1 < \alpha_2 < \ldots < \alpha_j} |x_{\alpha_1,n} x_{\alpha_2,n} \ldots x_{\alpha_j,n}|$$

from which the positivity of the $\{a_{j,n}\}$ follows easily. If one knows the asymptotic distribution of the zeros $\{x_{j,n}\}$ then one might wonder how the Taylor coefficients $\{a_{j,n}\}$ behave asymptotically. Under the condition that all the zeros of $\{\hat{p}_n(x)\}$ are negative one has the following result.

Theorem 5.1. (Van Assche-Fano-Ortolani [193]). Suppose that all the zeros of $\{\hat{p}_n(x) : n = 1,2,\ldots\}$ are in $[-A,0]$ $(0 < A)$ and suppose that ν_n converges weakly to a measure ν which has no mass at $\{0\}$, then there exists a concave and differentiable function g on (0,1) such that

$$\lim_{\substack{n \to \infty \\ j/n \to d}} \frac{1}{n} \log a_{j,n} = g(d) \tag{5.1.3}$$

for every $d \in (0,1)$. The function g satisfies

$$g'(d) = \log h^{-1}(d) \tag{5.1.4}$$

and

$$g(d) = -(1-d)\log h^{-1}(d) + \int_{-A}^{0} \log \{h^{-1}(d) - y\}\, d\nu(y) \tag{5.1.5}$$

for every $0 < d < 1$, where h^{-1} is the inverse function of

$$h(x) = - \int_{-A}^{0} \frac{y}{x-y}\, d\nu(y) \qquad x > 0 . \tag{5.1.6}$$

Sketch of the proof : this proof uses techniques from [55],[56] and [60]. We will only consider the special case where $\hat{p}_n(x) = \prod_{j=1}^{n} (x - y_{j,n})$ where

$$y_{j,n} = \nu^{-1}(\tfrac{j}{n}) \qquad j = 1,2,\ldots,n$$

with the inverse function ν^{-1} defined as

$$\nu^{-1}(y) = \inf\{x \in \mathbb{R} : \nu(x) \geq y\} ,$$

which makes ν^{-1} a left continuous increasing function on $(0,1]$. By construction these zeros satisfy the conditions of Theorem 5.1. The Taylor coefficients $\{a_{j,n}\}$ for these polynomials have the interesting property

$$(5.1.7) \qquad a_{k,n+m} \geq \sum_{i+j=k} a_{i,n} a_{j,m} .$$

This follows since for every positive integer α there exist integers α' and β' such that

$$\frac{\alpha' - 1}{n} \leq \frac{\alpha}{n + m} < \frac{\alpha'}{n}$$

$$\frac{\beta' - 1}{m} \leq \frac{\alpha}{n + m} < \frac{\beta'}{m} .$$

If we change every number $\frac{\alpha}{n+m}$ in the sequence $\left(\frac{\alpha_1}{n+m}, \frac{\alpha_2}{n+m}, \ldots, \frac{\alpha_k}{n+m}\right)$ by $\frac{\alpha'}{n}$ (if $\frac{\beta'}{m} > \frac{\alpha'}{n}$) or by $\frac{\beta'}{m}$ (if $\frac{\alpha'}{n} > \frac{\beta'}{m}$) then from the fact that $|\nu^{-1}|$ is decreasing we find

$$a_{k,n+m} = \frac{1}{k!} \sum_{\alpha_1 \neq \ldots \neq \alpha_k} \left|\nu^{-1}\left(\frac{\alpha_1}{n+m}\right) \ldots \nu^{-1}\left(\frac{\alpha_k}{n+m}\right)\right|$$

$$\geq \frac{1}{k!} \sum_{i=0}^{k} \binom{k}{i} \sum_{\alpha_1' \neq \ldots \neq \alpha_i'} \left|\nu^{-1}\left(\frac{\alpha_1'}{n}\right) \ldots \nu^{-1}\left(\frac{\alpha_i'}{n}\right)\right|$$

$$\times \sum_{\beta_1' \neq \ldots \neq \beta_{k-i}'} \left|\nu^{-1}\left(\frac{\beta_1'}{m}\right) \ldots \nu^{-1}\left(\frac{\beta_{k-i}'}{m}\right)\right|$$

$$= \sum_{i+j=k} \frac{1}{i!} \sum_{\alpha_1' \neq \ldots \neq \alpha_i'} |y_{\alpha_1',n} \cdots y_{\alpha_i',n}|$$

$$\times \frac{1}{j!} \sum_{\beta_1' \neq \ldots \neq \beta_j'} |y_{\beta_1',m} \cdots y_{\beta_j',m}|$$

which gives (5.1.7). If we introduce the continuous functions

$$g(d,n) = \begin{cases} \frac{1}{n} \log a_{j,n} & d = \frac{j}{n} \\ g(\frac{j}{n}, n) + (nd - j)\left\{g(\frac{j+1}{n}, n) - g(\frac{j}{n}, n)\right\} & \frac{j}{n} < d < \frac{j+1}{n} \end{cases}$$

then, for n fixed, $g(d,n)$ is a function in $[0,1]$ that connects the points $\left\{\frac{1}{n} \log a_{j,n} : j = 0,1,\ldots,n\right\}$ linearly. By (5.1.7) we see that

$$g(d, 2^{M+N+1}) \geq g(d, 2^{M+N})$$

whenever $d = K2^{-N}$ (K and N fixed). This means that $\{g(d,2^n)\}$ is increasing for n large enough whenever $d \in D = \{j2^{-n} : 0 \leq j \leq 2^n;\ n = 1,2,\ldots\}$ and therefore

$$\lim_{n \to \infty} g(d,2^n) = g(d) \qquad d \in D .$$

By (5.1.7) one finds

$$g\left(\frac{d_1 + d_2}{2}\right) \geq \frac{1}{2} \{g(d_1) + g(d_2)\} \qquad d_1, d_2 \in D$$

which makes g a concave function on D. Since D is dense in $[0,1]$ we can extend g to be a function on $(0,1)$ and by the continuity

$$\lim_{n \to \infty} g(d,2^n) = g(d) \qquad 0 < d < 1$$

and the convergence is uniform on closed intervals of $(0,1)$ ([83],[164],[165] Thm 7.13). Some extra manipulation is needed to go from the subsequence $\{2^n : n = 1,2,\ldots$ to the whole sequence of positive integers, to find

$$\lim_{n \to \infty} g(d,n) = g(d) \qquad 0 < d < 1 \quad ,$$

(see Fisher [60]). Since g is concave it will be differentiable on $(0,1)$ except possibly on a denumerable set $\Lambda \subset [0,1]$. The derivative g' is a decreasing function for which the limits from the left hand side and the right hand side exist for points in Λ; if we denote these limits by g'_- and g'_+ then $g'_-(d) > g'_+(d)$ $(d \in \Lambda)$. Define

$$\text{(5.1.9)} \qquad f(d) = e^{g'(d)} \qquad d \in (0,1)\setminus\Lambda$$

then f is decreasing with jumps at the points of Λ. The inverse function $h = f^{-1}$ exists on $(0,\infty)$ and is increasing and therefore differentiable except possibly for a denumerable set Λ^*. If $y \notin \Lambda^*$ then $h(y) \in (0,1)\setminus\Lambda$ so that g will be differentiable at $h(y)$. If x is fixed then $g'(d) - \log x$ is, by definition, equal to zero whenever $d = h(x)$, and therefore

$$(5.1.10) \qquad \sup_{0<d<1} \{g(d) + (1-d)\log x\} = g(h(x)) + \{1 - h(x)\} \log x$$

whenever $x \notin \Lambda^*$. This is true because $g(d) + (1-d)\log x$ has at most one maximum. It is clear that for every $j \leq n$

$$a_{j,n}x^{n-j} \leq \hat{p}_n(x) \qquad\qquad x > 0$$

from which

$$\frac{1}{n}\log a_{j,n} + \frac{n-j}{n}\log x \leq \frac{1}{n}\log \hat{p}_n(x) \ .$$

If we let n tend to infinity and j/n to $d \in (0,1)$ then by the uniform convergence and the weak convergence of ν_n to ν

$$\sup_{0<d<1} \{g(d) + (1-d)\log x\} \leq \int_{-A}^{0} \log(x-y)d\ (y) \ .$$

On the other hand

$$\hat{p}_n(x) \leq n \max_{0\leq j\leq n} a_{j,n}x^{n-j} \qquad\qquad x > 0$$

so that

$$\frac{1}{n}\log \hat{p}_n(x) \leq \frac{1}{n}\log n + \max_{0\leq j\leq n}\left\{\frac{1}{n}\log a_{j,n} + \frac{n-j}{n}\log x\right\} .$$

By taking limits we find

$$\int_{-A}^{0} \log(x-y)d\nu(y) \leq \sup_{0<d<1} \{g(d) + (1-d)\log x\} \ .$$

Combining both inequalities gives (5.1.5). Set $d = h(x)$ in (5.1.5) and differentiate with respect to x then (5.1.4) leads to (5.1.6). ∎

An immediate consequence of Theorem 5.1 is the following result :

Theorem 5.2. (Van Assche-Fano-Ortolani [193]) : With the same conditions and notation as in Thm 5.1 we have

(5.1.11) $$\lim_{\substack{n\to\infty \\ j/n\to d}} \frac{a_{j,n}}{a_{j-1,n}} = e^{g'(d)} = h^{-1}(d) \qquad 0 < d < 1$$

Proof : The Taylor coefficients of a polynomial $\hat{p}_n$ with only real zeros satisfy

$$(n-j+1)a_{j+1,n}a_{j-1,n} \le (n-j)a_{j,n}^2 \qquad j = 0,\dots,n$$

where $a_{-1,n} = a_{n+1,n} = 0$ ([30], thm 2.82). This implies

$$\frac{a_{j+1,n}}{a_{j,n}} < \frac{a_{j,n}}{a_{j-1,n}} \qquad j = 1,\dots,n$$

from which

$$\frac{a_{j+k,n}}{a_{j,n}} < \left(\frac{a_{j,n}}{a_{j-1,n}}\right)^k \qquad j = 1,\dots,n-k+1 .$$

Taking the kth root gives

$$\frac{a_{j,n}}{a_{j-1,n}} > \left(\frac{a_{j+k,n}}{a_{j,n}}\right)^{1/k} = \exp\left\{\frac{n}{k}\left(\frac{1}{n}\log a_{j+k,n} - \frac{1}{n}\log a_{j,n}\right)\right\} .$$

Let $k = [\varepsilon n]$, where $\varepsilon > 0$ is arbitrary small, and let n tend to infinity and j/n to d, then

$$\liminf \frac{a_{j,n}}{a_{j-1,n}} \ge \exp\left\{\frac{g(d+\varepsilon) - g(d)}{\varepsilon}\right\} .$$

In a similar way, using $a_{j,n}/a_{j-k,n} > (a_{j,n}/a_{j-1,n})^k$ one finds

$$\limsup \frac{a_{j,n}}{a_{j-1,n}} \le \exp\left\{\frac{g(d) - g(d-\varepsilon)}{\varepsilon}\right\} .$$

Let ε tend to zero to find the desired result.

∎

Example 1 (degenerate case) : Let $\{\hat{p}_n(x) : n = 1,2,...\}$ be a sequence of monic orthogonal polynomials with a spectral measure μ having a compact and denumerable support on the negative real line, with one accumulation point at $-a$ $(a > 0)$. From Lemma 1.4 in Chapter 1, with $E = \{-a\}$, we find that the sequence $\{\nu_n : n = 1,2,...\}$ converges weakly to the degenerate measure with all mass at the point $-a$, so that for $x > 0$

(5.1.12) $$\lim_{n \to \infty} \hat{p}_n(x)^{1/n} = x + a = \exp \int_{-\infty}^{0} \log(x-y)d\nu(y)$$

with ν the discrete measure with all mass at $-a$. From (5.1.6) we have

$$h(x) = \frac{a}{x+a} \qquad x > 0$$

so that

(5.1.13) $$h^{-1}(d) = a\,\frac{1-d}{d} \qquad 0 < d < 1 .$$

The function g in (5.1.5) becomes

$$g(d) = d \log a - d \log d - (1-d)\log(1-d) ,$$

which means that the coefficients of the polynomials $\{p_n(x) : n = 1,2,...\}$ have the asymptotic behaviour

(5.1.14) $$\begin{cases} \lim\limits_{\substack{n \to \infty \\ j/n \to d}} \frac{1}{n} \log a_{j,n} = d \log a - d \log d - (1-d)\log (1-d) \\ \\ \lim\limits_{\substack{n \to \infty \\ j/n \to d}} \frac{a_{j,n}}{a_{j-1,n}} = a\,\frac{1-d}{d} , \end{cases}$$

for every $0 < d < 1$. This result could also be obtained directly by using the invariance : the sequence $\{(x+a)^n : n = 1,2,...\}$ has the same asymptotic behaviour as in (5.1.12) and explicitely

$$(x + a)^n = \sum_{j=0}^{n} \binom{n}{j} a^j x^{n-j}$$

for which we find $a_{j,n} = \binom{n}{j}a^j$. The result in (5.1.14) then follows by using Stirling's formula.

Example 2 (orthogonal polynomials on a compact set) : Let $\{\hat{p}_n(x) : n = 0,1,2,\ldots\}$ be a sequence of monic orthogonal polynomials with a spectral measure μ which has its support on a compact set in $(-\infty,0]$ with positive capacity. If $\mathrm{supp}(\mu) = E \cup E^*$ and if μ has an absolute continuous part with respect to the Frostman measure μ_E (for which $\mathrm{supp}(\mu_E) = E$) with a Radon-Nikodym derivative w that is positive μ_E-almost everywhere, while E^* is a denumerable set with accumulation points in E then from Theorem 1.2 in Chapter 1 we find that ν_n converges weakly to the Frostman measure μ_E. Therefore

$$\begin{cases} \lim\limits_{\substack{n\to\infty \\ j/n\to d}} \frac{1}{n} \log a_{j,n} = g(d) \\ \\ \lim\limits_{\substack{n\to\infty \\ j/n\to d}} \frac{a_{j,n}}{a_{j-1,n}} = h^{-1}(d) \end{cases}$$

where

$$(5.1.15)\qquad \begin{cases} h(x) = -\int \frac{y}{x-y}\, d\mu_E(y) & x > 0 \\ \\ g(d) = -(1-d)\log h^{-1}(d) + \int_E \log\{h^{-1}(d) - y\}\, d\mu_E(y) & 0 < d < 1 . \end{cases}$$

For the asymptotic behaviour of the coefficients $\{a^*_{j,n} : j = 0,\ldots,n\}$ of the normalized orthogonal polynomials $\{p_n(x) : n = 0,1,2,\ldots\}$, given by $a^*_{j,n} = k_n a_{j,n}$, we have by (1.2.1)

$$\lim_{\substack{n\to\infty \\ j/n\to d}} \frac{1}{n} \log a^*_{j,n} = g(d) + \log \frac{1}{C(E)}$$

where $\log \frac{1}{C(E)}$ is Robin's constant and C(E) is the capacity of E.

If we take $E = [-1,0]$ then from (1.3.1) we find

$$\mu_E(A) = \frac{1}{\pi} \int_A \frac{dt}{\sqrt{t(1+t)}}$$

for every Borel set A in [-1,0]. Explicit computation gives

$$\int_E \log(x-y)\, d\mu_E(y) = -2 \log 2 + \log \left\{2x + 1 + 2\sqrt{x^2 + x}\right\} \qquad x > 0$$

and by taking the derivative with respect to the variable x

$$h(x) = 1 - x \int_E \frac{1}{x-y}\, d\mu_E(y) = 1 - \frac{x}{\sqrt{x^2+x}} \qquad x > 0$$

which gives

$$h^{-1}(d) = \frac{(1-d)^2}{1-(1-d)^2} \qquad 0 < d < 1 .$$

If we use (5.1.5) we also find

$$g(d) = -2 \log 2 - d \log d + (2-d)\log (2-d) - 2(1-d)\log (1-d) \qquad 0<d<1$$

so that the asymptotic behaviour of the coefficients of the monic polynomials is given by

$$(5.1.16) \quad \begin{cases} \lim\limits_{\substack{n\to\infty \\ j/n\to d}} \frac{1}{n} \log a_{j,n} = -2 \log 2 - d \log d + (2-d)\log (2-d) - 2(1-d)\log (1-d) \\[2ex] \lim\limits_{\substack{n\to\infty \\ j/n\to d}} \frac{a_{j,n}}{a_{j-1,n}} = \frac{(1-d)^2}{1-(1-d)^2} \end{cases}$$

for $0 < d < 1$. The result in (5.1.15) could be obtained in a more direct way by means of the invariance : the polynomials $\{P_n^{(\alpha,\beta)}(1+2x) : n = 0,1,2,\ldots\}$, with $P_n^{(\alpha,\beta)}(x)$ the Jacobi polynomial with parameters α and β ($\alpha,\beta > -1$), are orthogonal on $[-1,0]$ with weight function

$$w(x;\alpha,\beta) = (-x)^\alpha(1+x)^\beta$$

and satisfy the conditions needed to use Theorem 1.2 (with $E = [-1,0]$). The coefficients can be found from (0.1.6) giving

$$a_{j,n} = \frac{\Gamma(\alpha+n+1)}{\Gamma(\alpha+\beta+2n+1)} \binom{n}{j} \frac{\Gamma(\alpha+\beta+2n-j+1)}{\Gamma(\alpha+n-j+1)} .$$

Stirling's formula then gives (5.1.15). We have plotted the functions g and h^{-1} for $E = [-1,0]$ in figures 5.1 and 5.2.

<u>Example 3</u> (orthogonal polynomials with exponential weights) : If $\{\hat{p}_n(x) : n=0,1,2,\ldots\}$ are orthogonal polynomials on $(-\infty,0]$ with a weight function w that is almost everywhere positive on $(-\infty,0]$ and for which

$$\lim_{x \to -\infty} |x|^{-\gamma} \log w(x) = -1 \qquad \gamma > 0$$

then by Theorem 4.3 (for $(-\infty,0]$) we have

$$\lim_{n \to \infty} \frac{1}{n} \log \{c_n^{-n} \hat{p}_n(c_n x)\} = \int_{-1}^{0} \log(x-y) d\mu^{\gamma^-}(y)$$

where

$$c_n = \left(\frac{4n}{\lambda_{2\gamma}}\right)^{1/\gamma} \qquad ; \qquad \lambda_{2\gamma} = \frac{2\Gamma(\gamma + \frac{1}{2})}{\sqrt{\pi}\,\Gamma(\gamma)}$$

and μ^{γ^-} is the Ullman measure on $[-1,0]$ with parameter γ :

$$\mu^{\gamma^-}(A) = \int_A b_\gamma(x)dx$$

with A a Borel set in $[-1,0]$ and

$$b_\gamma(x) = \frac{1}{\sqrt{|x|}} v(2\gamma;\sqrt{|x|}) \qquad -1 < x < 0 .$$

If

$$\hat{p}_n(x) = \sum_{j=0}^{n} a_{j,n} x^{n-j}$$

then

$$c_n^{-n} \hat{p}_n(c_n x) = \sum_{j=0}^{n} a_{j,n}\, c_n^{-j}\, x^{n-j}$$

so that

$$\begin{cases} \lim\limits_{\substack{n \to \infty \\ j/n \to d}} \{\frac{1}{n} \log a_{j,n} - d \log c_n\} = g(d) \\ \\ \lim\limits_{\substack{n \to \infty \\ j/n \to d}} \frac{1}{c_n} \frac{a_{j,n}}{a_{j-1,n}} = h^{-1}(d) \end{cases}$$

where

$$h(x) = -\int_{-1}^{0} \frac{y}{x-y} b_\gamma(y)\, dy \qquad x > 0$$

$$g(d) = -(1-d)\log h^{-1}(d) + \int_{-1}^{0} \log\{h^{-1}(d) - y\}\, b_\gamma(y)dy \qquad 0 < d < 1 .$$

For $\gamma = 1$ we have

$$b_1(x) = \frac{2}{\pi}\frac{\sqrt{1+x}}{\sqrt{-x}} \qquad -1 < x < 0$$

and with some calculus

$$\int_{-1}^{0} b_1(y)\log(x-y)dy = 2\log(x + \sqrt{x^2+x}) - \log x + 2(-x + \sqrt{x^2+x})$$
$$-2\log 2 - 1 \qquad x > 0 .$$

Differentiating with respect to x gives

$$h(x) = 1 - \frac{2x}{x + \sqrt{x^2+x}} = 2x + 1 - 2\sqrt{x^2+x} \qquad x > 0 ,$$

and after inversion

$$h^{-1}(d) = \frac{(1-d)^2}{4d} .$$

The function g is given by

$$g(d) = -2(1-d)\log(1-d) - d\log d - 2d\log 2 - d .$$

Moreover $c_n = 4n$, so that

$$(5.1.17) \quad \begin{cases} \lim\limits_{\substack{n\to\infty \\ j/n\to d}} \{\frac{1}{n}\log a_{j,n} - d\log n\} = -2(1-d)\log(1-d) - d\log d - d \\ \lim\limits_{\substack{n\to\infty \\ j/n\to d}} \frac{1}{4n}\frac{a_{j,n}}{a_{j-1,n}} = \frac{(1-d)^2}{4d} \end{cases}$$

for $0 < d < 1$. This result can also be found directly by the invariance : the sequence $\{L_n^{(\alpha)}(-x) : n = 0,1,2,\ldots\}$, with $L_n^{(\alpha)}(x)$ the Laguerre polynomial with parameter $\alpha > -1$, satisfies the needed conditions and from (0.1.9) we find

$$a_{j,n} = \binom{n+\alpha}{j} \frac{n!}{(n-j)!}$$

which, by means of Stirling's formula, gives (5.1.16). We have plotted the right hand sides of (5.1.17) in figures 5.1 and 5.2.

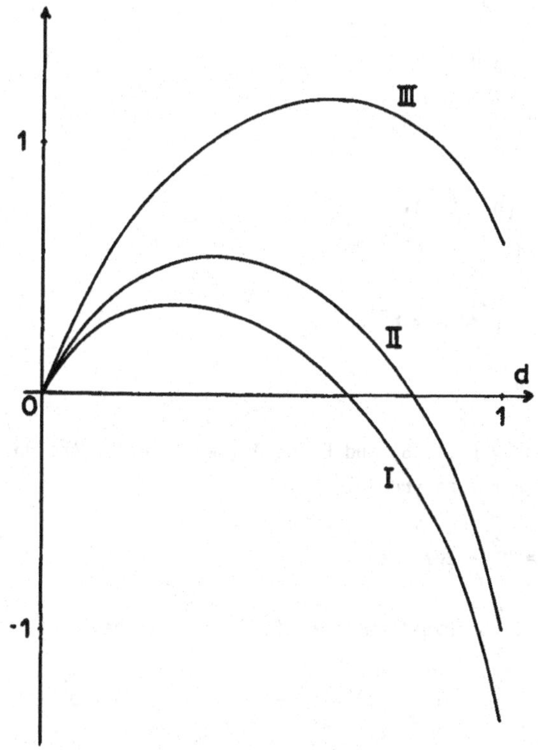

fig. 5.1 : the function g for
I : orthogonal polynomials on [-1,0]
II : Laguerre polynomials on $(-\infty,0]$
III : iterations of $z^2 + 6z + 1$.

Example 4 (iterations of a polynomial) : Let T be a polynomial of degree $k \geq 2$,

$$T(z) = z^k + a_{k-1} z^{k-1} + \dots + a_0$$

and consider the iterations $\{T_n(z) : n = 0,1,2,\dots\}$, with $T_0(z) = z$ and $T_n(z) = T_{n-1}(T(z))$. The relation with the orthogonal polynomials $\{p_n(x) : n =$

$0,1,2,\ldots\}$ corresponding to the equilibrium measure μ_J on the Julia set J of T is

$$\hat{p}_{k^n}(x) = T_n(x) + \frac{1}{k} a_{k-1}$$

(Chapter 1, § 1.4). We can apply Theorem 5.1 and 5.2 to the coefficients of $\{p_n(x) : n = 0,1,2,\ldots\}$ provided the Julia set J is in $(-\infty,0]$. Let

$$T_n(x) = \sum_{j=0}^{k^n} t_{j,n}\, x^{k^n-j}$$

then for Julia sets $J \subset (-\infty,0]$

$$\begin{cases} \lim\limits_{\substack{n\to\infty \\ j/k^n \to d}} \frac{1}{k^n} \log t_{j,n} = g(d) \\[2ex] \lim\limits_{\substack{n\to\infty \\ j/k^n \to d}} \frac{t_{j,n}}{t_{j-1,n}} = h^{-1}(d) \end{cases}$$

with g and h given by (5.1.15) and E the Julia set J for the polynomial T. As an example we consider the polynomial

$$T(z) = z^2 + 2bz + c\ .$$

By means of the Möbius transformation $L(z) = z - b$ we have

$$\tilde{T}(z) = L^{-1} \circ T \circ L(z) = z^2 - p \quad , \qquad p = b^2 - b - c\ .$$

We know that the Julia set $\tilde{J}$ for $z^2 - p$ $(p > 2)$ is a real set in $[-a,a]$, where $a = \frac{1}{2} + \sqrt{\frac{1}{4} + p}$ and for $p > 2$ this is a Cantor set. The Julia set J for T is then given by $L(\tilde{J})$ which means that $J \subset [-a-b,a-b]$ whenever $b^2 - b - c > 2$. If we choose the parameters b and c such that

$$b > 0\ ,\ c > 0\ ,\ b^2 - b - c > 2$$

then $J \subset (-\infty,0]$. We have plotted the functions g and h^{-1} for $T(z) = z^2 + 6z + 1$ in figures 5.1 and 5.2. These functions were determined numerically by explicit calculation of the coefficients of $T_{10}(z)$, which is a polynomial of degree 1024.

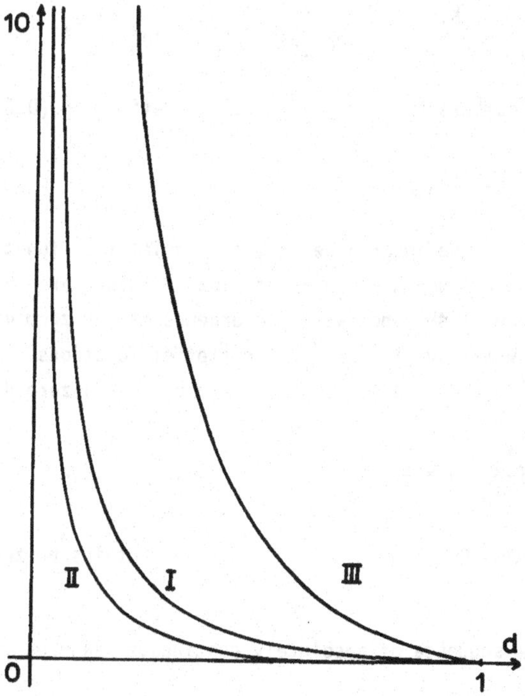

fig. 5.2. : the function h^{-1} for

I : orthogonal polynomials on [-1,0]

II : Laguerre polynomials on (-∞,0]

III : iterations of $z^2 + 6z + 1$.

5.2. Weak asymptotics for Christoffel functions

Let $\{p_n(x) : n = 0,1,2,...\}$ be a sequence of orthogonal polynomials, then the *Christoffel functions* $\{\lambda_n(x) : n = 1,2,...\}$ are defined as

$$\lambda_n(x) = \{\sum_{j=0}^{n-1} p_j^2(x)\}^{-1} . \tag{5.2.1}$$

Clearly $1/\lambda_n(x)$ is a polynomial of degree 2n-2 which is positive for $x \in \mathbb{R}$ and by the Christoffel-Darboux formula (0.2.16)

$$\lambda_n(x) = \frac{k_n}{k_{n-1}} \{p_n'(x)p_{n-1}(x) - p_{n-1}'(x)p_n(x)\}^{-1} .$$

The Christoffel numbers $\{\lambda_{j,n} : j = 1,\ldots,n\}$, defined in (0.2.2), are given by

$$\lambda_{j,n} = \lambda_n(x_{j,n})$$

which already shows the importance of these Christoffel functions. Christoffel functions and their importance in approximation theory are thoroughly discussed in Nevai [151]. We will show how weak convergence of the zero distributions immediately implies some weak asymptotics for Christoffel functions.

Let $\{\nu_n : n = 1,2,\ldots\}$ be the (contracted) zero distribution measures

$$(5.2.2) \qquad \begin{cases} \nu_n(\{\frac{x_{j,n}}{c_n}\}) = \frac{1}{n} & j = 1,\ldots,n \\ \nu_n(A) = 0 & A \text{ contains no zeros of } p_n(c_n x) \end{cases}$$

and define a new sequence of absolutely continuous measures $\{\gamma_n : n = 1,2,\ldots\}$ by

$$(5.2.3) \qquad \gamma_n(A) = \frac{1}{n} \int_A \frac{1}{\lambda_n(c_n x)} \, d\mu(c_n x)$$

where μ is the spectral measure for the orthogonal polynomials $\{p_n(x) : n = 0,1,2,\ldots\}$, then

Theorem 5.3. (Nevai [138] p. 49; Van Assche [192]) : Suppose that ν_n converges weakly to a probability measure ν and suppose that $\frac{1}{c_n} \frac{k_{n-1}}{k_n}$ is of the order $o(\sqrt{n})$, then γ_n converges weakly to ν.

Proof : Recall the fundamental polynomials of Lagrange interpolation

$$(5.2.3) \qquad L_{k,n}(x) = \frac{k_{n-1}}{k_n} \lambda_{k,n} \, p_{n-1}(x_{k,n}) \cdot \frac{p_n(x)}{x - x_{k,n}}$$

By means of the Christoffel-Darboux formula (0.2.15) one easily obtains

$$(5.2.4) \qquad L_{k,n}(x) = \lambda_{k,n} \sum_{j=0}^{n-1} p_j(x) p_j(x_{k,n})$$

By (5.2.4) and the Gauss-Jacobi quadrature

$$\sum_{k=1}^{n} \frac{1}{\lambda_{k,n}} L_{k,n}^2(x) = \sum_{j=0}^{n-1} p_j^2(x)$$

so that for every bounded continuous function f

$$\left|\int_{-\infty}^{\infty} f(x)d\nu_n(x) - \int_{-\infty}^{\infty} f(x)d\gamma_n(x)\right|$$

$$= \left|\frac{1}{n}\sum_{j=1}^{n} f\left(\frac{x_{j,n}}{c_n}\right) - \frac{1}{n}\sum_{j=1}^{n} \frac{1}{\lambda_{j,n}} \int_{-\infty}^{\infty} f(x)L_{j,n}^2(c_nx)d\mu(c_nx)\right|$$

$$< \frac{1}{n}\sum_{j=1}^{n}\frac{1}{\lambda_{j,n}}\left\{\int_{|x-\frac{x_{j,n}}{c_n}|<\varepsilon} + \int_{|x-\frac{x_{j,n}}{c_n}|\geq\varepsilon} \left|f\left(\frac{x_{j,n}}{c_n}\right) - f(x)\right| L_{j,n}^2(c_nx)d\mu(c_nx)\right\}$$

$$= I_1 + I_2$$

where $\varepsilon > 0$ is arbitrary. Easy estimation gives

$$I_1 = \frac{1}{n}\sum_{j=1}^{n}\frac{1}{\lambda_{j,n}} \int_{|x-\frac{x_{j,n}}{c_n}|<\varepsilon} \left|f\left(\frac{x_{j,n}}{c_n}\right) - f(x)\right| L_{j,n}^2(c_nx)d\mu(c_nx)$$

$$< \omega_f(\varepsilon)\frac{1}{n}\sum_{j=1}^{n}\frac{1}{\lambda_{j,n}}\int_{-\infty}^{\infty} L_{j,n}^2(x)d\mu(x)$$

$$= \omega_f(\varepsilon)$$

where

$$\omega_f(\varepsilon) = \sup_{|x-y|<\varepsilon} |f(x) - f(y)|$$

is the modulus of continuity of f. Also, by (5.2.3)

$$I_2 = \frac{1}{n}\sum_{j=1}^{n}\frac{1}{\lambda_{j,n}} \int_{|x-\frac{x_{j,n}}{c_n}|\geq\varepsilon} \left|f\left(\frac{x_{j,n}}{c_n}\right) - f(x)\right| L_{j,n}^2(c_nx)d\mu(c_nx)$$

$$< 2\,\|f\|_\infty \left(\frac{k_{n-1}}{k_n}\right)^2 \frac{1}{n}\sum_{j=1}^{n} \lambda_{j,n}\, p_{n-1}^2(x_{j,n}) \int_{|x-x_{j,n}|\geq\varepsilon c_n} \frac{p_n^2(x)}{(x-x_{j,n})^2} d\mu(x)$$

$$< \frac{2\|f\|_\infty}{\varepsilon^2} \frac{1}{n} \left(\frac{1}{c_n} \frac{k_{n-1}}{k_n}\right)^2 \sum_{j=1}^{n} \lambda_{j,n} p_{n-1}^2(x_{j,n})$$

$$= \frac{2\|f\|_\infty}{\varepsilon^2} \frac{1}{n} \left(\frac{1}{c_n} \frac{k_{n-1}}{k_n}\right)^2 .$$

By the conditions imposed I_2 tends to zero as n tends to infinity and the result follows since ε is arbitrary and $\omega_f(\varepsilon) \longrightarrow 0$ as $\varepsilon \longrightarrow 0$.

∎

The extra condition that $\frac{1}{c_n} \frac{k_{n-1}}{k_n} = o(\sqrt{n})$ is automatically fulfilled if the support of the measure ν is compact. Indeed, by taking $f(x) = |x|^M$ we have by the weak convergence of ν_n

$$\frac{1}{n} \sum_{j=1}^{n} \left|\frac{x_{j,n}}{c_n}\right|^M < c_M$$

where c_M is some constant that does not depend on n. Clearly

$$\frac{1}{n} \left|\frac{x_{k,n}}{c_n}\right|^M < c_M$$

and therefore $\dfrac{\max_{1\le k\le n} |x_{k,n}|}{c_n} = O(n^{1/M})$ for every M and this implies that $\frac{1}{c_n} \frac{k_{n-1}}{k_n} = o(\sqrt{n})$.

This Theorem gives weak asymptotics for Christoffel functions which can be applied to all the examples treated in the previous section.

5.3. Toeplitz matrices associated with orthogonal polynomials

Let $\{p_n(x) : n = 0,1,2,\ldots\}$ be orthogonal polynomials with spectral measure μ and let g be a real-valued μ-measurable function such that for every positive integer k

$$\int_{-\infty}^{\infty} |x^k g(x)| \, d\mu(x) < \infty$$

then the infinite *Toeplitz matrix* $T(g;\mu)$ associated with the orthogonal polynomials $\{p_n(x) : n = 0,1,2,\ldots\}$ consists of the entries

(5.3.1) $$(T(g;\mu))_{i,j} = \int_{-\infty}^{+\infty} p_i(x)p_j(x)g(x)d\mu(x) \qquad i,j = 0,1,2,\ldots$$

The truncated matrix $T_n(g;\mu)$ consists of the first n rows and columns of $T(g;\mu)$:

(5.3.2) $$T_n(g;\mu) = (T(g;\mu))_{i,j=0,1,\ldots,n-1} \cdot$$

For orthogonal polynomials on unbounded sets we can modify these notions. If $\{c_n : n = 1,2,\ldots\}$ is the contraction sequence then we can define

(5.3.3) $$T_n^*(g;\mu) = \left[\int_{-\infty}^{\infty} p_i(x)p_j(x)g(\frac{x}{c_n})d\mu(x)\right]_{i,j=0,1,\ldots,n-1} .$$

If $c_n = 1$ for every n then (5.3.3) and (5.3.2) coincide. Notice that

$$\text{Trace } T_n^*(g;\mu) = \int_{-\infty}^{\infty} \frac{1}{\lambda_n(c_nx)} g(x)\ d\mu(c_nx) = n \int_{-\infty}^{\infty} g(x)\ d\gamma_n(x)$$

where $\lambda_n(x)$ is the Christoffel function (5.2.1) and γ_n is the measure given in (5.2.3). We already know by Theorem 5.3 that γ_n converges weakly to the asymptotic zero distribution (if this exists). We will also need the weak convergence of a bivariate measure. Define a sequence of measures $\{\xi_n : n = 1,2,\ldots\}$ on $\mathbb{R}^2$ by

(5.3.4) $$\xi_n(A) = \frac{1}{n} \iint_A K_n^2(c_nx,c_ny)d\mu(c_nx)d\mu(c_ny)$$

where A is a Borel set in $\mathbb{R}^2$ and

(5.3.5) $$K_n(x,y) = \sum_{j=0}^{n-1} p_j(x)p_j(y)$$

then we have the following result :

<u>Lemma 5.4.</u> (Nevai [144]; Van Assche [192]) : Suppose that ν_n in (5.2.2) converges weakly to some probability measure ν and suppose that $\frac{1}{c_n}\frac{k_{n-1}}{k_n} = o(\sqrt{n})$, then ξ_n converges weakly to a measure ξ which is concentrated on the diagonal $\{(x,x) : x \in \mathbb{R}\}$ and for every bounded continuous function $f(x,y)$

(5.3.6) $$\lim_{n\to\infty} \int_{-\infty}^{\infty}\int_{-\infty}^{\infty} f(x,y)d\xi_n(x,y) = \int_{-\infty}^{\infty} f(x,x)d\nu(x) .$$

Proof : In order to prove (5.3.6) we restrict ourselves to functions of the form $f(x)g(y)$ with f and g bounded and continuous functions with compact support. Clearly

$$|\int_{-\infty}^{\infty}\int_{-\infty}^{\infty} f(x)g(y)d\xi_n(x,y) - \int_{-\infty}^{\infty} f(x)g(x)d\gamma_n(x)|$$

$$= \frac{1}{n} |\int_{-\infty}^{\infty}\int_{-\infty}^{\infty} K_n^2(c_nx,c_ny)\{f(x)g(y) - f(x)g(x)\} d\mu(c_nx)d\mu(c_ny)\}|$$

$$< \frac{1}{n}\int_{-\infty}^{\infty} |f(y)|\, d\mu(c_ny)\{\int_{|x-y|<\varepsilon} + \int_{|x-y|>\varepsilon} |g(x)-g(y)|K_n^2(c_nx,c_ny)d\mu(c_nx)$$

$$= I_1 + I_2 .$$

We easily estimate

$$I_1 < \|f\|_\infty\, \omega_g(\varepsilon)\, \frac{1}{n} \int_{-\infty}^{\infty}\int_{-\infty}^{\infty} K_n^2(x,y)d\mu(x)d\mu(y)$$

$$= \|f\|_\infty\, \omega_g(\varepsilon)$$

and by the Christoffel-Darboux formula (0.2.15)

$$I_2 = \frac{1}{n}\left(\frac{1}{c_n}\frac{k_{n-1}}{k_n}\right)^2 \int_{-\infty}^{\infty} |f(x)|\, d\mu(c_nx)$$

$$\times \int_{|x-y|>\varepsilon} |g(x)-g(y)| \frac{\{p_{n-1}(c_nx)p_n(c_ny)-p_n(c_nx)p_{n-1}(c_ny)\}^2}{(x-y)^2}\, d\mu(c_ny)$$

$$< 2\|f\|_\infty \|g\|_\infty \frac{1}{n}\left(\frac{1}{c_n}\frac{k_{n-1}}{k_n}\right)^2 \frac{1}{\varepsilon^2}$$

$$\times \int_{-\infty}^{\infty}\int_{-\infty}^{\infty} \{p_{n-1}^2(x)p_n^2(y)+p_n^2(x)p_{n-1}^2(y)-2p_{n-1}(x)p_n(y)p_n(x)p_{n-1}(y)\}d\mu(x)d\mu(y)$$

$$= \frac{4}{\varepsilon^2}\|f\|_\infty \|g\|_\infty \frac{1}{n}\left(\frac{1}{c_n}\frac{k_{n-1}}{k_n}\right)^2 .$$

By our conditions I_2 tends to zero as $n \to \infty$ and I_1 can be made as small as possible by choosing ε small. The result then follows by Theorem 5.3.

■

Denote the eigenvalues of $T_n^*(g;\mu)$ by $\Lambda_{1,n} < \Lambda_{2,n} < \ldots < \Lambda_{n,n}$. Of course $T_n^*(g;\mu)$ is real and symmetric so that these eigenvalues are also real. Introduce another sequence of measures ρ_n by

$$(5.3.7)\qquad \begin{cases} \rho_n(\{\Lambda_{j,n}\}) = \frac{k}{n} & \text{if } \Lambda_{j,n} \text{ has multiplicity } k \\ \rho_n(A) = 0 & \text{if } A \text{ contains no eigenvalues} \end{cases}$$

then ρ_n describes the distribution of the eigenvalues of the Toeplitz matrix $T_n^*(g;\mu)$. Notice that for $g(x) = x$ the measure ρ_n coincides with ν_n. We can now state and prove the following theorem on the distribution of the eigenvalues of Toeplitz matrices associated with orthogonal polynomials :

Theorem 5.5. (Nevai [144]; Van Assche [192]) : Suppose that the measure ν_n given in (5.2.2) converges weakly to some probability measure ν and that $\frac{1}{c_n}\frac{k_{n-1}}{k_n} = o(\sqrt{n})$. If g is a bounded measurable function with discontinuities on a set of ν-measure zero, then ρ_n, given by (5.3.7), converges weakly to νg^{-1} for which

$$\nu g^{-1}(A) = \nu(g^{-1}(A))$$

for every Borel set A.

Proof : Let G be a bounded continuous function. Define polynomials $\varphi_{k,n}(x)$ by

$$\varphi_{k,n}(x) = \sum_{j=0}^{n-1} b_{j,k}\, p_j(x)$$

where $\{b_k = (b_{0,k}, b_{1,k}, \ldots, b_{n-1,k}) : k = 1,2,\ldots,n\}$ is a collection of orthonormalized eigenvectors of $T_n^*(g;\mu)$, that is

$$T_n(g;\mu) b_k' = \Lambda_{k,n}\, b_k'$$

$$b_k \cdot b_\ell' = \delta_{k,\ell}$$

then

$$(5.3.8)\qquad \int_{-\infty}^{\infty} \varphi_{k,n}(x)\varphi_{\ell,n}(x)\,d\mu(x) = \delta_{k,\ell}$$

$$(5.3.9)\qquad \int_{-\infty}^{\infty} \varphi_{k,n}(x)\varphi_{\ell,n}(x) g(x)\,d\mu(x) = \Lambda_{k,n}\,\delta_{k,\ell}$$

and

(5.3.10) $$\sum_{k=1}^{n} \varphi_{k,n}(x)\varphi_{k,n}(y) = K_n(x,y) .$$

Clearly by (5.3.10)

$$\frac{1}{n}\sum_{k=1}^{n} G(\Lambda_{k,n}) - \frac{1}{n}\int_{-\infty}^{\infty} G(g(x))K_n(c_nx,c_nx)d\mu(c_nx)$$

$$= \frac{1}{n}\sum_{k=1}^{n}\int_{-\infty}^{\infty}\left\{G(\Lambda_k,n) - G(g(x))\right\}\varphi_{k,n}^2(c_nx)d\mu(c_nx) .$$

Fix $\varepsilon > 0$ and choose $\delta > 0$ so that $|G(x) - G(y)| < \varepsilon$ for $|x-y| < \delta$, then

$$\int_{-\infty}^{\infty} G(x)d\rho_n(x) - \int_{-\infty}^{\infty} G(g(x))d\gamma_n(x)$$

$$= \frac{1}{n}\sum_{k=1}^{n}\left(\int_{|\Lambda_{k,n}-g(x)|<\delta} + \int_{|\Lambda_{k,n}-g(x)|\geq\delta}\right)\left\{G(\Lambda_{k,n})-G(g(x))\right\}\varphi_{k,n}^2(c_nx)d\mu(c_nx).$$

$$= I_1 + I_2 .$$

Then

$$|I_1| < \varepsilon\,\frac{1}{n}\sum_{k=1}^{n}\int_{-\infty}^{\infty}\varphi_{k,n}^2(x)d\mu(x) = \varepsilon$$

and

$$|I_2| < \frac{2}{\delta^2}\,\|G\|_\infty\,\frac{1}{n}\sum_{k=1}^{n}\int_{-\infty}^{\infty}\{\Lambda_{k,n} - g(x)\}^2\,\varphi_{k,n}^2(c_nx)d\mu(c_nx)$$

$$= \frac{2}{\delta^2}\,\|G\|_\infty\,\frac{1}{n}\sum_{k=1}^{n}\int_{-\infty}^{\infty}\{\Lambda_{k,n}^2 - 2g(x)\Lambda_{k,n} + g^2(x)\}\varphi_{k,n}^2(c_nx)d\mu(c_nx) .$$

Use (5.3.8) - (5.3.10) to find

(5.3.11) $$|I_2| < \frac{2}{\delta^2}\,\|G\|_\infty\,\frac{1}{n}\left\{\int_{-\infty}^{\infty} g^2(x)K_n(c_nx,c_nx)d\mu(c_nx) - \sum_{k=1}^{n}\Lambda_{k,n}^2\right\} .$$

One can check that

$$K_n(x,x) = \int_{-\infty}^{\infty} K_n^2(x,y)d\mu(y)$$

and

$$\sum_{k=1}^{n} \Lambda_{k,n}^2 = \text{Trace}(T_n^*(g;\mu))^2 = \int_{-\infty}^{\infty}\int_{-\infty}^{\infty} g(x)g(y)K_n^2(c_nx,c_ny)d\mu(c_nx)d\mu(c_ny)$$

so that (5.3.11) can be rewritten as

$$|I_2| < \frac{2}{\delta^2} \|G\|_\infty \frac{1}{n} \int_{-\infty}^{\infty}\int_{-\infty}^{\infty} \{g^2(x) - g(x)g(y)\}K_n^2(c_nx,c_ny)d\mu(c_nx)d\mu(c_ny) .$$

Then use Lemma 5.4 to find

$$\lim_{n\to\infty} I_2 = 0$$

and since ε is arbitrary the result is therefore proven. ∎

Theorem 5.5 was first considered by Grenander and Szegö [82] for the case $c_n = 1$ for every n and they call the measure ν a canonical distribution. These authors formulate the Theorem for the case where supp(μ) = [-1,1] but, as indicated by Nevai ([138], p. 49) their proof is not valid. The more general statement given in this section is proved by using an idea of Nevai [144]. For more literature regarding Toeplitz matrices associated with orthogonal polynomials we refer to Grenander-Szegö [82], Nevai [138]; [144], [151],Máté-Nevai-Totik [120].

5.4. Functions of the second kind

Let $\{p_n(x) : n = 0,1,2,...\}$ be orthogonal polynomials with spectral measure μ. Functions of the second kind $\{q_n(x) : n = 0,1,2,...\}$ are defined by the integral

$$(5.4.1) \qquad q_n(x) = \int_{-\infty}^{\infty} \frac{p_n(y)}{x - y}\, d\mu(y) .$$

These functions are well defined whenever $x \in \mathbb{C}\backslash\text{supp}(\mu)$. A straightforward analysis shows that $\{q_n(x) : n = 0,1,2,...\}$ satisfy the same recurrence relation as the polynomials $\{p_n(x) : n = 0,1,2,...\}$ but with initial conditions

$$(5.4.2) \qquad a_0q_{-1}(x) = 1 \quad , \quad q_0(x) = \int_{-\infty}^{\infty} \frac{d\mu(y)}{x - y} .$$

Observe that q_0 is the Stieltjes transform of the spectral measure μ. Functions of the second kind are usually only studied for classical orthogonal polynomials (Szegö [175]) but recently more general orthogonal polynomials have also been considered (Barrett [16], Case [34], Geronimo [71]).

If we rewrite the definition (5.4.1) somewhat we have

$$p_n(x)q_n(x) = \int_{-\infty}^{\infty} \frac{p_n(x) - p_n(y)}{x - y} p_n(y)d\mu(y) + \int_{-\infty}^{\infty} \frac{p_n^2(y)}{x-y} d\mu(y) .$$

Clearly $\{p_n(x) - p_n(y)\}/(x-y)$ is a polynomial in the variable y of degree n-1 and therefore by orthogonality

$$p_n(x)q_n(x) = \int_{-\infty}^{\infty} \frac{p_n^2(y)}{x - y} d\mu(y) \tag{5.4.3}$$

which means that the product $p_n(x)q_n(x)$ is the Stieltjes transform of a measure which is absolutely continuous with respect to the spectral measure μ with Radon-Nikodym derivative $p_n^2(x)$. In particular

$$\frac{c_n}{n} \sum_{j=0}^{n-1} p_j(c_n x)q_j(c_n x) = S(\gamma_n;x)$$

where γ_n is the measure defined in (5.2.3). This leads immediately to the following result :

Theorem 5.6. : Suppose that the measure ν_n in (5.2.2) converges weakly to some probability measure ν and that $\frac{1}{c_n} \frac{k_{n-1}}{k_n} = o(\sqrt{n})$, then

$$\lim_{n \to \infty} \frac{c_n}{n} \sum_{j=0}^{n-1} p_j(c_n x)q_j(c_n x) = S(\nu;x)$$

uniformly on compact subsets of $\mathbb{C}\setminus\mathbb{R}$.

In the spectral case $c_n = 1$ $(n = 1,2,\ldots)$ this gives the (C,1) convergence of the sequence $\{p_n(x)q_n(x) : n = 1,2,3,\ldots\}$ outside the spectrum. This can be improved to ordinary convergence when the recurrence coefficients behave nicely :

Theorem 5.7. : (Rakhmanov [161]) Suppose that the recurrence coefficients a_n and b_n satisfy

$$\lim_{n\to\infty} a_n = a > 0 \quad , \quad \lim_{n\to\infty} b_n = b ,$$

then

$$\lim_{n\to\infty} p_n(x)q_n(x) = \frac{1}{\sqrt{(x-b)^2 - 4a^2}}$$

uniformly on compact subsets of $\mathbb{C}\setminus\text{supp}(\mu)$. As a consequence

$$\lim_{n\to\infty} \int_{-\infty}^{\infty} f(y)p_n^2(y)d\mu(y) = \frac{1}{\pi} \int_{b-2a}^{b+2a} \frac{f(y)}{\sqrt{4a^2 - (y-b)^2}} dy$$

for every continuous function f.

<u>Proof</u> : Since $\{q_n(x)\}$ is a solution of the same recurrence relations as the one for $\{p_n(x)\}$, we can look for the Wronskian $W(q_n,p_n)$, which gives

$$a_{n+1}\{q_n(x)p_{n+1}(x) - q_{n+1}(x)p_n(x)\} = 1$$

from which one easily finds

$$a_{n+1} \left\{ \frac{p_{n+1}(x)}{p_n(x)} - \frac{q_{n+1}(x)}{q_n(x)} \right\} = \frac{1}{q_n(x)p_n(x)} .$$

By Poincaré's result [157] the ratio's $p_{n+1}(x)/p_n(x)$ and $q_{n+1}(x)/q_n(x)$ converge to one of the solutions of the quadratic equation

$$ag^2(x) - (x-b)g(x) + a = 0$$

provided that $|g(x)| \neq 1$. Clearly

$$\lim_{x\to\infty} \frac{p_{n+1}(x)}{xp_n(x)} = a_{n+1}$$

$$\lim_{x\to\infty} \frac{q_{n+1}(x)}{q_n(x)} \frac{p_{n+1}(x)}{p_n(x)} = \lim_{x\to\infty} \frac{x\int_{-\infty}^{\infty} \frac{p_{n+1}^2(y)}{x-y} d\mu(y)}{x\int_{-\infty}^{\infty} \frac{p_n^2(y)}{x-y} d\mu(y)} = 1$$

so that

$$\lim_{n \to \infty} \frac{p_{n+1}(x)}{p_n(x)} = \frac{x - b + \sqrt{(x-b)^2 - 4a^2}}{2a}$$

$$\lim_{n \to \infty} \frac{q_{n+1}(x)}{q_n(x)} = \frac{x - b - \sqrt{(x-b)^2 - 4a^2}}{2a}$$

and therefore

$$\lim_{n \to \infty} q_n(x)p_n(x) = \frac{1}{\sqrt{(x-b)^2 - 4a^2}} .$$

This convergence is uniform on compact subsets of $\mathbb{C}\backslash\text{supp}(\mu)$ since on any compact set $K \subset \mathbb{C}\backslash\text{supp}(\mu)$ we have

$$|q_n(x)p_n(x)| < \int_{-\infty}^{\infty} \frac{p_n^2(y)}{|x-y|} d\mu(y) < \frac{1}{\delta}$$

where δ is the distance from K to supp(μ), so that the Stieltjes-Vitali theorem applies. ∎

The above result shows that under the appropriate conditions the functions of the second kind $q_n(x)$ behave as $1/p_n(x)$ outside the spectrum. An immediate consequence of Theorem 5.8 is

<u>Theorem 5.9</u> (Barrett [16]) : Suppose that supp(μ) = [-1,1] and that

$$\int_{-1}^{1} \frac{\log \mu'(y)}{\sqrt{1 - y^2}} dy > -\infty$$

then

$$\lim_{n \to \infty} q_n(x) \cdot (x + \sqrt{x^2 - 1})^n = \frac{\sqrt{2\pi}}{\sqrt{x^2 - 1}} D(x - \sqrt{x^2 - 1})$$

uniformly on compact subsets of $\mathbb{C}\backslash[-1,1]$, where D(z) is Szegö's function

$$D(z) = \exp \frac{1}{4\pi} \int_{-\pi}^{\pi} \log[\mu'(\cos\theta)|\sin\theta|] \frac{1 + ze^{-i\theta}}{1 - ze^{-i\theta}} d\theta$$

Proof : In this case the recurrence coefficients will satisfy

$$\lim_{n \to \infty} a_n = \frac{1}{2} \qquad , \qquad \lim_{n \to \infty} b_n = 0$$

and

$$\lim_{n \to \infty} p_n(x)(x + \sqrt{x^2 - 1})^{-n} = \left\{\sqrt{2\pi}\, D(x - \sqrt{x^2 - 1})\right\}^{-1}$$

uniformly on compact sets of $\mathbb{C}\backslash[-1,1]$ (see § 1.3.1 in Chapter 1).

∎

CHAPTER 6 : SOME APPLICATIONS

In this Chapter we give some applications of the theory of orthogonal polynomials on the real line and of the results obtained in the previous chapters. The most obvious applications are to be found in numerical analysis where the zeros of orthogonal polynomials are considered to be interesting nodes for interpolation and numerical quadrature. Orthogonal polynomials are also frequently used in the theory of Fourier series when one wants to approximate a certain function on the real line in some Hilbert space by means of an orthogonal series. We will not discuss these applications but refer to the existing literature (e.g. [54],[61],[138],[175]). We will concentrate on some less well known applications : the distribution of eigenvalues of some random $n \times n$ matrices, discrete Schrödinger equations in one dimension and some stochastic processes known as birth-and-death processes.
Of course there are still more applications like time series analysis [82], neutron transport [87] and quantum chemistry [200].

6.1. Random matrices

Some orthogonal polynomials play an important role if one wants to find the distribution of eigenvalues of random matrices. Random matrices are being used in statistics and in physics. They were introduced in statistics by Wishart (1928) and Hsu (1939) and enter in a natural way if one wants to estimate the covariance matrix of a random vector. A review of random matrices in statistics is given by James [91]. Around 1952 Wigner suggested to use random matrices in physics for studying the energy levels for n-particle systems. If the number of particles is large then it is impossible to calculate the interaction $H_{i,j}$ between two particles i and j ($i \neq j$) or the potential $H_{i,i}$ of particle i because there is so much interference of all the other particles. For that reason one assumes that all the entries $\{H_{i,j} : i,j = 1,2,\ldots,n\}$ of the matrix H are random variables defined on the same probability space $(\Omega, \mathcal{B}, P)$. The matrix H is an operator from $\mathbb{C}^n$ to $\mathbb{C}^n$ and is called the *Hamiltonian* for this n-particle system. The eigenvalues of H correspond to the possible values of the energy for a particle so that there is a lot of interest to obtain the distribution of the eigenvalues. The most relevant references for random matrices are Mehta [130] and Wigner [203]. Mehta [130] refers briefly to the connection with orthogonal polynomials.

We only consider Hermitian random matrices. Let $(\Omega,\mathcal{B},P)$ be a probability space in which n^2 random variables $\{X_{a,b} : 1 \leq a \leq b \leq n\}$ and $\{Y_{a,b} : 1 \leq a < b \leq n\}$ are defined and let H_n be a random matrix for which

$$(H_n)_{a,a} = X_{a,a} \qquad a = 1,2,\ldots,n$$

$$\left.\begin{aligned}(H_n)_{a,b} &= X_{a,b} + i\,Y_{a,b}\\ (H_n)_{b,a} &= X_{a,b} - i\,Y_{a,b}\end{aligned}\right\} \qquad 1 \leq a < b \leq n\,.$$

The joint distribution function for this random matrix is

$$P(H_n \leq x) = P(X_{a,a} \leq x_{a,a}\,,\ X_{a,b} \leq x_{a,b}\,,\ Y_{a,b} \leq y_{a,b} : 1 \leq a < b \leq n)$$

where x is an Hermitian matrix with entries $\{x_{a,b} + iy_{a,b} : 1 \leq a \leq b \leq n\}$. We will assume that the joint distribution function is given by

$$\text{(6.1.1)} \qquad P(H_n \leq x) = C_n \int_{t\leq x} \exp\{-\mathrm{Tr}\ Q(t)\}dt$$

where t is an Hermitian $n \times n$-matrix, Q is a polynomial of degree 2M, tr A is the trace of the matrix A and C_n is a normalizing constant. By means of a one-to-one relation between the entries $\{X_{a,b} : 1 \leq a \leq b \leq n\} \cup \{Y_{a,b} : 1 \leq a < b \leq n\}$ and $\{\Lambda_1,\ldots,\Lambda_n\} \cup \{\Gamma_i : i = 1,2,\ldots,n^2-n\}$, where $\{\Lambda_i\}$ are the eigenvalues of H_n and $\{\Gamma_i\}$ are n^2-n additional parameters needed to determine the matrix H_n completely (e.g. n^2-n linearly independent components of the eigenvectors of H_n), we obtain

$$P(\Lambda_a \leq \lambda_a : a = 1,\ldots,n;\ \Gamma_b \leq \gamma_b : 1 \leq b \leq n^2 - n)$$

$$= C_n \int_{u\leq\lambda} \int_{v\leq\gamma} \exp\{-\sum_{i=1}^{n} Q(u_i)\}|J_n(u,v)|dv\,du$$

with J_n the Jacobian of the transformation :

$$J_n(u,v) = \frac{\partial(x_{11},\ldots,x_{n,n},x_{12},y_{12},\ldots,x_{n-1,n},y_{n-1,n})}{\partial(u_1,\ldots,u_n,v_1,\ldots,v_{n^2-n})}\,.$$

One can work out this Jacobian (Mehta [130], p. 37) and find

$$|J_n(u,v)| = \prod_{i<j} (u_i - u_j)^2\ f(v)$$

where f is a function which only depends on v and not on u. If we integrate over the variables v then we find the joint distribution function of the eigenvalues

$$P(\Lambda_1 < \lambda_1, \ldots, \Lambda_n < \lambda_n)$$

$$= C_n' \int_{-\infty}^{\lambda_1} \cdots \int_{-\infty}^{\lambda_n} \exp\{-\sum_{i=1}^{n} Q(u_i)\} \prod_{i<j} (u_i - u_j)^2 \, du_1 \ldots du_n$$

where C_n' is another normalizing constant. Notice that the random variables $\{\Lambda_1, \ldots, \Lambda_n\}$ are exchangeable so that every Λ_i has the same marginal distribution. In order to obtain the distribution of Λ_1 we will integrate over the variables $\{u_2, \ldots, u_n\}$. In *Vandermonde's determinant*

$$\prod_{i<j} (x_j - x_i) = \begin{vmatrix} 1 & 1 & \cdots & 1 \\ x_1 & x_2 & \cdots & x_n \\ x_1^2 & x_2^2 & \cdots & x_n^2 \\ \vdots & \vdots & & \vdots \\ x_1^{n-1} & x_2^{n-1} & & x_n^{n-1} \end{vmatrix}$$

we are allowed to add to each row a linear combination of all the other rows. If $\{p_n(x) : n = 0,1,2,\ldots\}$ are the orthogonal polynomials with weight function $w(x) = \exp\{-Q(x)\}$, then consequently

$$\prod_{i<j} (x_j - x_i) = \left(\prod_{m=1}^{n-1} \frac{1}{k_m} \right) \det (p_{i-1}(x_j))$$

and therefore

$$P(\Lambda_1 < \lambda_1, \ldots, \Lambda_n < \lambda_n)$$

$$= C_n'' \int_{-\infty}^{\lambda_1} \cdots \int_{-\infty}^{\lambda_n} \exp\left\{-\sum_{j=1}^{n} Q(u_j)\right\} \{\det(p_{i-1}(u_j))\}^2 \, du_1 \ldots du_n .$$

The determinant of $(p_{i-1}(u_j))$ is given by

$$\det(p_{i-1}(u_j)) = \sum_{\sigma} (-1)^{|\sigma|} \, p_{j_1-1}(u_1) \cdots p_{j_n-1}(u_n) ,$$

the sum runs over all permutations $\sigma = (j_1, j_2, \ldots, j_n)$ of $(1,2,\ldots,n)$ and $|\sigma|$ is the order of the permutation σ. Therefore

$$\{\det(p_{i-1}(u_j))\}^2 = \sum_{\sigma}\sum_{\tau} (-1)^{|\sigma|+|\tau|} \, p_{j_1-1}(u_1)p_{i_1-1}(u_1)$$

$$\cdots p_{j_n-1}(u_n)p_{i_n-1}(u_n) .$$

If we integrate over the variables $(u_2,\ldots,u_n)$ then by orthogonality only the terms for which $i_k = j_k$ will have a non-zero contribution and thus

(6.1.2) $$P(\Lambda_1 < \lambda_1) = \frac{1}{n} \int_{-\infty}^{\lambda_1} e^{-Q(u)} \sum_{j=0}^{n-1} p_j^2(u) \, du .$$

This means that every eigenvalue has a density $f_n(x)$ given by

$$f_n(x) = \frac{1}{n} \sum_{j=0}^{n-1} p_j^2(x) \, e^{-Q(x)}$$

where $\{p_n(x) : n = 0,1,2,\ldots\}$ are orthogonal polynomials with weight function $e^{-Q(x)}$. For every n one can therefore obtain the distribution of the eigenvalues in terms of a special system of orthogonal polynomials.

As a consequence of Theorem 5.3 and Theorem 4.3 we find that for each eigenvalue $\Lambda^{(n)}$ of the random $n \times n$ matrix H_n with Q an even polynomial of degree 2M and leading coefficient 1

(6.1.3) $$\left(\frac{\lambda_{2M}}{2n}\right)^{1/2M} \Lambda^{(n)} \xrightarrow{\mathcal{D}} \Lambda$$

where Λ is a random variable distributed according to the Ullman measure μ^{2M} and $\mathcal{D}$ denotes convergence in distribution. The special case $Q(x) = x^2$ is well known and gives the so-called *semicircle law of Wigner* (Arnold [4] - [8], Mehta [130], Marchenko-Pastur [115], Wigner [202] - [203]). The case $Q(x) = x^4 + ax^2$ has been considered by Bessis [23] and a general outline for Q an even polynomial was given by Bessis-Itzykson-Zuber [25] where techniques from quantum theory were used.

In a similar way one can calculate the marginal distributions of a random vector

$(\Lambda_1,\ldots,\Lambda_n)$ with joint probability distribution

$$P(\Lambda_1 \leqslant \lambda_1,\ldots,\Lambda_n \leqslant \lambda_n) = C_n' \int_{-\infty}^{\lambda_1} \ldots \int_{-\infty}^{\lambda_n} \prod_{i<j} (u_i - u_j)^2 \, d\mu(u_1)\ldots d\mu(u_n)$$

where μ is some probability measure on the real line for which all the moments exist. If $\{p_n(x;\mu) : n = 0,1,2,\ldots\}$ are the orthogonal polynomials with spectral measure μ then

$$P(\Lambda_1 \leqslant \lambda_1) = \int_{-\infty}^{\lambda_1} \frac{1}{n} \sum_{j=0}^{n-1} p_j^2(x;\mu) \, d\mu(x) \ .$$

Theorem 5.3 then gives the asymptotic behaviour of each eigenvalue $\Lambda^{(n)}$ of the n-vector $(\Lambda_1,\ldots,\Lambda_n)$ as n tends to infinity.

6.2. Discrete Schrödinger equations

Schrödinger's equation is one of the most important equations in quantum theory. If a particle is able to move in one dimension then this equation is

$$(6.2.1) \qquad - \frac{\hbar^2}{2m} \frac{d^2\psi(t)}{dt^2} + V(t)\psi(t) = E\psi(t)$$

with m the mass of the particle, $\hbar$ is Planck's constant, $V(t)$ the potentiel energy of a particle in position t and E the total energy of the particle. Equation (6.2.1) implies an eigenvalue problem for the operator

$$H = - \frac{\hbar^2}{2m} \frac{d^2}{dt^2} + V$$

with domain in a subset of the square integrable functions on $(-\infty,\infty)$. The eigenvalues of this operator are the possible values that the energy can have. A (normalized) eigenfunction ψ corresponding to the eigenvalue E is called a *wave function* (for the energy E). The function $|\psi(t)|^2$ is a probability density function and the probability to find a particle with energy E in (a,b) is given by

$$\int_a^b |\psi(t)|^2 \, dt \ .$$

We will now approximate the differential equation (6.2.1) by a difference equation (Case-Kac [35]). Since

$$\lim_{\varepsilon \to 0} \frac{1}{\varepsilon^2} \{\psi(t+\varepsilon) - 2\psi(t) + \psi(t-\varepsilon)\} = \frac{d^2\psi(t)}{dt^2}$$

we find that the recurrence relation

$$-\frac{\hbar^2}{2m} \{\psi(E,(n+1)\Delta) - 2\psi(E,n\Delta) + \psi(E,(n-1)\Delta)\}/\Delta^2$$

$$+ V(n\Delta)\psi(E,n\Delta) = E\psi(E,n\Delta)$$

goes over to (6.2.1) if $\Delta \longrightarrow 0$ and $n\Delta \longrightarrow t$. Putting $\Delta^2 E = -x$, $\psi(E,n\Delta) = p_n(x)$ and $\Delta^2 V(n\Delta) = v_n$ gives

$$\text{(6.2.2)} \qquad xp_n(x) = \frac{\hbar^2}{2m} p_{n+1}(x) - (\frac{\hbar^2}{m} + v_n)p_n(x) + \frac{\hbar^2}{2m} p_{n-1}(x)$$

and this is a recurrence relation of the same type as the recurrence relation for orthogonal polynomials, with

$$a_n = \frac{\hbar^2}{2m} \quad , \quad b_n = -(\frac{\hbar^2}{m} + v_n) .$$

The coefficients $\{v_n : n = 0,1,2,\ldots\}$ correspond to the (discrete) potential of the particle. Another way of discretizing the Schrödinger equation uses the result

$$\lim_{\varepsilon \to 0} \frac{1}{\varepsilon^2} \{\psi(t+\varepsilon) + \psi(t-\varepsilon) - 2e^{\varepsilon^2 q(t)} \psi(t)\} = \frac{d^2\psi(t)}{dt^2} - 2q(t)\psi(t)$$

so that

$$-\frac{\hbar}{2m} \{\psi(E,(n+1)\Delta) + \psi(E,(n-1)\Delta)\} = (-\frac{\hbar^2}{m} + E\Delta^2)e^{\Delta^2 V(n\Delta)m/\hbar^2} \psi(E,n\Delta)$$

tends to (6.2.1) if $\Delta \longrightarrow 0$. Setting $\frac{\hbar^2}{m} - E^2\Delta^2 = x$, $\Delta^2 V(n\Delta)m/\hbar = v_n$ and $\exp(v_n/2)\psi(E,n\Delta) = p_n(x)$ gives

$$\text{(6.2.3)} \qquad xp_n(x) = \frac{\hbar^2}{2m} \left\{\exp(-\frac{v_n + v_{n+1}}{2}) p_{n+1}(x) + \exp(-\frac{v_n + v_{n-1}}{2}) p_{n-1}(x)\right\}$$

and this is again just like the recurrence relation for orthogonal polynomials with

$$a_n = \frac{\hbar}{2m} \exp\left(-\frac{v_n + v_{n-1}}{2}\right) \quad , \quad b_n = 0$$

This means that the recurrence relation for orthogonal polynomials can be interpreted as a discrete version of Schrödinger's equation. The corresponding operator H is the Jacobi matrix J and, depending on ones preference, either the truncated Jacobi matrix J_n or the infinite Jacobi matrix J can be considered (e.g. Dehesa [48]). If one considers the truncated Jacobi matrix J_n then the eigenvalues are the zeros of the polynomial $p_n(x)$ which give the possible values for the total energy in this model. The eigenvector for $x_{j,n}$ is given by $\sqrt{\lambda_{j,n}}\,(p_0(x_{j,n}),p_1(x_{j,n}),\ldots,p_{n-1}(x_{j,n}))$ and an arbitrary particle with energy $x_{j,n}$ can be at n places with a probability that it is at place k given by $\lambda_{j,n}p^2_{k-1}(x_{j,n})$. If one considers the infinite Jacobi matrix J_∞ then the eigenvalues are the discrete mass points x_m of the spectral measure μ for the orthogonal polynomials. The corresponding eigenvector is

$$\frac{1}{\left(\sum_{j=0}^{\infty} p_j^2(x_m)\right)^{1/2}} (p_0(x_m),p_1(x_m),\ldots) \ .$$

The discrete eigenvalues are called *bound states* for a particle in this model and the probability that a particle, in the bound state x_m is at position k is given by

$$\frac{1}{\sum_{j=0}^{\infty} p_j^2(x_m)} p_{k-1}^2(x_m) \ .$$

That part of supp(μ) where the continuous part of spectral measure is strictly increasing is called the *continuous spectrum*.

A discrete Schrödinger equation which has been studied a lot is *Harper's equation*

$$xp_n(x) = p_{n+1}(x) + 2\cos(2\pi n\alpha - \nu)p_n(x) + p_{n-1}(x)$$

where α and ν are two parameters (Hofstadter [86]). If α is a rational number then $\{\cos(2\pi n\alpha - \nu) : n = 0,1,2,\ldots\}$ is periodic and therefore one can use the results in Chapter 2 to conclude that the continuous spectrum consists of several disjoint intervals and there are only a finite number of bound states with at most one bound state between two consecutive intervals of the continuous spectrum. If α is not a rational number then $\{\cos(2\pi n\alpha - \nu) : n = 0,1,2,\ldots\}$ is almost-periodic and the results from Chapter 2 do not apply any longer. A special case of almost-periodic sequences are limit-periodic sequences $\{R_n : n = 0,1,2,\ldots\}$ for which there exists a sequence $\{\lambda_n : n = 1,2,\ldots\}$ such that

$$\lim_{n \to \infty} |R_{\lambda_n + s} - R_s| = 0$$

uniformly in s. The recurrence coefficients for orthogonal polynomials on a Julia set corresponding to the iterations of $z^2 - p$ $(p > 3)$ are limit-periodic by (1.4.16). The study of such polynomials therefore gives some insight in the chaotic behaviour of some almost-periodic Schrödinger operators ([11],[21],[136]). The continuous spectrum for orthogonal polynomials on the Julia set of $z^2 - p$ $(p > 2)$ is a Cantor set, which gives an indication that physical phenomena for such operators may be very strange.

6.3. Birth-and-death processes

Karlin and McGregor ([94] - [99]) have showed in 1957 that orthogonal polynomials play a fundamental role for some stochastic processes. Orthogonal polynomials on $[0,\infty)$ are important in the theory of birth-and-death processes and orthogonal polynomials on $[-1,1]$ for random walks.

A stationary proces $\{X(t) : t \geq 0\}$ is a *birth-and-death process* if it is a stationary Markov process with values in the set $\mathbb{N}$ of non-negative integers with a transition matrix $\mathbb{P}(t)$ for which the entries

$$P_{n,m}(t) = P(X(t+s) = m | X(s) = n)$$

behave for $t \longrightarrow 0$ as

$$(6.3.1) \quad \begin{cases} P_{n,n+1}(t) = \lambda_n t + o(t) \\ P_{n,n-1}(t) = \mu_n t + o(t) \\ P_{n,n}(t) = 1 - (\lambda_n + \mu_n)t + o(t) \\ P_{n,m}(t) = o(t) & |m - n| > 1 . \end{cases}$$

The $\{\lambda_n : n = 0,1,2,\ldots\}$ are *birth rates* and $\{\mu_n : n = 0,1,2,\ldots\}$ are *death rates*. The transition matrix $\mathbb{P}$ satisfies the backward equation

$$\mathbb{P}'(t) = A\mathbb{P}(t)$$

where A is the infinitesimal operator

$$A = \lim_{t \to 0} \frac{\mathbb{P}(t) - I}{t} = \begin{bmatrix} -(\lambda_0 + \mu_0) & \lambda_0 & 0 & 0 & \\ \mu_1 & -(\lambda_1 + \mu_1) & \lambda_1 & 0 & \\ 0 & \mu_2 & -(\lambda_2 + \mu_2) & \lambda_2 & \cdots \\ \vdots & \ddots & \ddots & \ddots & \end{bmatrix}$$

If the o-terms in (6.3.1) are uniform in n and m then $\mathbb{P}$ also satisfies the forward equation

$$\mathbb{P}'(t) = \mathbb{P}(t)A .$$

If we suppose that $\lambda_n > 0$, $\mu_n > 0$ $(n = 1,2,\ldots)$, $\lambda_0 > 0$ and $\mu_0 \geq 0$ then we can define a sequence of polynomials $\{Q_n(x) : n = 0,1,\ldots\}$ by

$$-xQ_n(x) = \lambda_n Q_{n+1}(x) - (\lambda_n + \mu_n)Q_n(x) + \mu_n Q_{n-1}(x)$$

$$Q_0(x) = 1 \quad , \quad Q_{-1}(x) = 0$$

or in matrix notation $-xQ(x) = AQ(x)$. The recurrence relation for the monic polynomials $\hat{p}_n(x) = (-1)^n \lambda_0 \lambda_1 \ldots \lambda_{n-1} Q_n(x)$ becomes

$$\hat{p}_{n+1}(x) = \{x - (\lambda_n + \mu_n)\}\hat{p}_n(x) - \lambda_{n-1}\mu_n \hat{p}_{n-1}(x)$$

and by Favard's theorem (Lemma 0.3) we find that $\{\hat{p}_n(x) : n = 0,1,2,\ldots\}$ are orthogonal polynomials with some spectral measure μ on the real line. One can show that $\text{supp}(\mu) \subset [0,\infty)$. If we introduce the sequence $\{\pi_n : n = 0,1,2,\ldots\}$ by

$$\pi_n = \frac{\lambda_0\lambda_1 \cdots \lambda_{n-1}}{\mu_1\mu_2 \cdots \mu_n}$$

then the transition probabilities are explicitely given by

$$P_{ij}(t) = \pi_j \int_0^\infty e^{-xt}\, Q_i(x)\, Q_j(x)\, d\mu(x) \;. \tag{6.3.3}$$

In a similar way one can show that orthogonal polynomials on [-1,1] are important for a stationary Markov process $\{X_n : n = 0,1,\ldots\}$ with values in the set of non-negative integers, with a transition matrix $\mathbb{P}$ for which the entries

$$P_{n,m} = P(X_{k+1} = m \mid X_k = n)$$

satisfy

$$\begin{cases} P_{n,n-1} = q_n & \\ P_{n,n+1} = p_n & p_n + q_n + r_n \leq 1 \\ P_{n,n} = r_n & \\ P_{n,m} = 0 & |n-m| > 1 \;. \end{cases}$$

Such a process is often referred to as a *random walk*. If $p_n > 0$, $q_n > 0$ $(n = 1,2,\ldots)$, $p_0 > 0$ and $q_0 \geq 0$ then we define a sequence of polynomials $\{R_n(x) : n = 0,1,2,\ldots\}$ by

$$xR_n(x) = p_nR_{n+1}(x) + r_nR_n(x) + q_nR_{n-1}(x)$$

$$R_0(x) = 1 \quad , \quad R_{-1}(x) = 0 \;.$$

These polynomials are again orthogonal (but not necessarily normalized) with a spectral measure λ and now one can show that $\text{supp}(\lambda) \subset [-1,1]$. The transition probabilities now become

$$P_{i,j}^{(n)} = P(X_{k+n} = j \mid X_k = i)$$

$$= \Pi_j \int_{-1}^1 x^n\, R_i(x)R_j(x)d\lambda(x)$$

where

$$\pi_n = \frac{p_0 p_1 \cdots p_{n-1}}{q_1 \cdots q_n} .$$

In this theory it is very important to find the spectral measure μ or λ when the recurrence coefficients are given. For a queue M/M/s for which the time between two arrivals has an exponential distribution with parameter λ and for which there are s servers which all have a serving time which is exponentially distributed with parameter μ, the birth and death rates are

$$\lambda_n = \lambda \qquad n = 0,1,2,\ldots$$

$$\mu_n = \begin{cases} n\mu & n < s \\ s\mu & n \geq s \end{cases}$$

which means that the coefficients are constant from some subscript on. The spectral measure in that case is given by Theorem 2.25. If the birth and death rates are periodic then the spectral measure follows by Theorem 2.14. (Woess [205]). The behaviour of the spectral measure μ in the neighbourhood of 0 determines whether the birth-and-death process is transient or recurrent, while the behaviour of the spectral measure λ in the neighbourhood of 1 shows whether one deals with a transient or recurrent random walk. If the stochastic process is transient or positive recurrent then one can find the rate of convergence for the transition probabilities to their stationary values when there is an interval containing zero (or containing 1) on which μ (or λ) has no mass (except possibly at zero or one) (Callaert [33]).

APPENDIX

A.1. The Stieltjes transform

If μ is a measure (not necessarily a positive measure) on the real line, then its Stieltjes transform is given by

$$(A.1.1) \qquad S(\mu;z) = \int_{-\infty}^{\infty} \frac{d\mu(x)}{z-x} .$$

This function is analytic in $\mathbb{C}\setminus\mathrm{supp}(\mu)$. If $\mathrm{supp}(\mu) = \mathbb{R}$ then the Stieltjes transform is analytic in both $\{z \in \mathbb{C} : \mathrm{Im}\, z > 0\}$ and $\{z \in \mathbb{C} : \mathrm{Im}\, z < 0\}$. The Stieltjes transform $S(\mu;z)$ determines the measure μ uniquely (up to a constant). This follows from the inversion formula

$$(A.1.2) \qquad \frac{1}{2}\{\mu((-\infty,x]) + \mu((-\infty,x))\} - \frac{1}{2}\{\mu(-\infty,y]) + \mu((-\infty,y))\}$$

$$= \frac{1}{2\pi i} \lim_{v \to 0} \int_x^y \{S(\mu;u-iv) - S(\mu;u+iv)\}du .$$

(Wintner [204], p. 93-96). An important result concerning weak convergence of measures is the *theorem of Grommer and Hamburger* :

THEOREM A.1 (Wintner [204], p. 104-105) : Let $\{\mu_n : n = 1,2,\ldots\}$ be a sequence of measures in $\mathbb{R}$ for which the total variation is uniformly bounded.

(a) If $\mu_n \to \mu$ then $S(\mu_n;z) \longrightarrow S(\mu;z)$ uniformly on compact sets of $\mathbb{C}\setminus\mathbb{R}$.

(b) If $S(\mu_n;z) \longrightarrow S(z)$ uniformly on compact sets of $\mathbb{C}\setminus\mathbb{R}$, then $S(z)$ is the Stieltjes transform of a measure μ on $\mathbb{R}$ and $\mu_n \to \mu$.

This theorem shows that Stieltjes transforms behave better than Fourier transforms, as far as weak convergence is concerned, since the limit of a converging sequence of Stieltjes transforms is again a Stieltjes transform, which is in general not true for Fourier transforms.

Example 1 : If μ is a degenerate measure with mass 1 at the point $a \in \mathbb{R}$, then

$$S(\mu;z) = \frac{1}{z - a} .$$

In general if $S(\mu;z)$ is a rational function with real poles and positive residues, then μ is a positive measure with only mass points at those poles.

Example 2 : Suppose μ is the arcsin measure on $[-1,1]$ so that for every Borel set A in $[-1,1]$

$$\mu(A) = \frac{1}{\pi} \int_A \frac{dt}{\sqrt{1 - t^2}} ,$$

then

$$S(\mu;z) = \frac{1}{\sqrt{z^2 - 1}} \qquad z \in \mathbb{C}\backslash[-1,1]$$

where the square root is such that $|x + \sqrt{x^2 - 1}| > 1$ when $x \in \mathbb{C}\backslash[-1,1]$.

Example 3 : Suppose that μ is the measure on $[-1,1]$ such that

$$\mu(A) = \frac{2}{\pi} \int_A \sqrt{1 - t^2}\, dt \qquad z \in \mathbb{C}\backslash[-1,1]$$

for every Borel set A in $[-1,1]$, then

$$S(\mu;z) = \frac{2}{x + \sqrt{x^2 - 1}} = 2\{x - \sqrt{x^2 - 1}\}$$

where the square root is the same as in Example 2.

Example 4 : Let μ be the Cauchy measure on $\mathbb{R}$, i.e.

$$\mu(A) = \frac{1}{\pi} \int_A \frac{dt}{1 + t^2}$$

for every Borel set A in $\mathbb{R}$, then

$$S(\mu;z) = \begin{cases} \dfrac{1}{z + i} & \text{if } \operatorname{Im} z > 0 \\[2ex] \dfrac{1}{z - i} & \text{if } \operatorname{Im} z < 0 . \end{cases}$$

Example 5 : Let μ be the measure on $[0,\infty)$ such that

$$\mu(A) = \frac{1}{\pi}\int_A \frac{dt}{\sqrt{t}\,(1+t)}$$

for every Borel set A in $[0,\infty)$, then

$$S(\mu;z) = \frac{1}{1+z} - \frac{1}{\sqrt{-z}\,(1+z)} \qquad z \in \mathbb{C}\setminus[0,\infty)$$

where the square root $\sqrt{-z}$ is real and positive if $z < 0$.

A.2. The Stieltjes convolution

We will first introduce some notions from the theory of distributions. For more information we refer to Bremermann [31] and Rudin [166]. Let $C^\infty(\mathbb{R})$ be the space of all complex-valued functions with continuous derivatives of all orders. Let $\mathcal{D}_K$ be the space of all functions in C^∞ with support in the compact set $K \subset \mathbb{R}$. The union of all $\mathcal{D}_K$ is a vector space that we denote by $\mathcal{D}$ and which is the space of test functions. A sequence $\{\phi_j : j = 1,2,\ldots\}$ converges in $\mathcal{D}$ to ϕ if there exists a compact set K such that the support of ϕ_j is a subset of K for every j and if for every positive integer N

$$\text{(A.2.1)} \qquad \lim_{j\to\infty} \sup_{\substack{k \le N\\ x\in K}} |\phi_j^{(k)}(x) - \phi^{(k)}(x)| = 0$$

where $\phi^{(k)}$ is the kth derivative of ϕ. A linear functional Λ on $\mathcal{D}$ which is continuous with respect to this topology is a *(Schwarz) distribution*. The space of distributions is denoted by $\mathcal{D}'$. If N is the smallest integer such that for every compact set K there exists a constant C for which

$$\text{(A.2.2)} \qquad |\Lambda\phi| \le C\|\phi\|_N$$

for every $\phi \in \mathcal{D}$, with

$$\|\phi\|_N = \sup_{\substack{k \le N\\ x\in K}} |\phi^{(k)}(x)|$$

then N is the order of the distribution Λ. The distribution Λ_μ for which

$$(A.2.3) \qquad \Lambda_\mu \phi = \int \phi(x) d\mu(x) \qquad \phi \in \mathcal{D}$$

with μ a complex measure on $\mathbb{R}$, is usually identified with the measure μ. If μ is a degenerate measure with total mass 1 at the point a then the associated distribution is denoted by δ_a,

$$\delta_a \phi = \phi(a) \qquad \phi \in \mathcal{D}$$

and this is the so-called Dirac distribution at the point a. The space $\mathcal{O}_\alpha$ contains all functions in C^∞ for which

$$\phi^{(k)}(t) = O(|t|^\alpha) \qquad t \to \infty$$

for every integer k. A sequence $\{\phi_j : j = 1,2,\ldots\}$ converges in $\mathcal{O}_\alpha$ to ϕ if the sequence $\{\phi_j^{(k)} : j = 1,2,\ldots\}$ converges to $\phi^{(k)}$ uniformly on compact subsets of $\mathbb{R}$ and if there exists a constant C_k such that

$$|\phi_j^{(k)}(t)| < C_k |t|^\alpha$$

for every integer k. The space $\mathcal{O}'_\alpha$ contains all continuous linear functionals on $\mathcal{O}_\alpha$ and clearly $\mathcal{O}'_\alpha \subset \mathcal{D}'$. If Λ is a distribution in $\mathcal{D}'$ then its derivatives are defined by

$$\Lambda^{(k)}\phi = (-1)^k \Lambda \phi^{(k)} \qquad \phi \in \mathcal{D}$$

If there exists a distribution Γ such that $\Lambda\phi = \Gamma^{(1)}\phi$ then Γ is an anti derivative of Λ. Every distribution Λ in $\mathcal{D}'$ has an anti derivative which is unique up to a constant.

If Λ is a distribution in $\mathcal{O}'_\alpha$ $(\alpha > -1)$ then its Stieltjes transform is defined as

$$S(\Lambda;z) = \Lambda f_z \quad ; \quad f_z(x) = \frac{1}{z-x} \qquad (z \in \mathbb{C}\setminus\mathbb{R}) \, .$$

This definition is compatible with the definition in (A.1.1) and (A.2.3) because Λ_μ always belongs to $\mathcal{O}'_{-1}$. The inversion formula (A.1.2) now becomes

$$(A.2.4) \qquad \frac{1}{2\pi i} \lim_{v \to 0} \{S(\Lambda;x-iv) - S(\Lambda;x+iv)\}\phi(x)dx = \Lambda\phi \, .$$

We define the *Stieltjes convolution* of two complex measures μ and μ as the linear functional

$$(\mu\Delta\lambda)\phi = \iint \frac{\phi(x) - \phi(t)}{x - t} d\mu(x)d\lambda(t) \qquad \phi \in \mathcal{D}$$

This double integral is well defined since the integral is a continuous function of the two variables x and t provided we define it as $\phi'(t)$ if $t = x$. By the mean value theorem we know that $\phi(x) - \phi(t) = \phi'(y)(x - t)$, where y is between x and t, so that

$$|(\mu\Delta\lambda)| < \iint \frac{|\phi(x) - \phi(t)|}{|x - t|} d|\mu|(x)\, d|\lambda|(t)$$

$$< \|\mu\|\|\lambda\| \sup_{x \in \mathbb{R}} |\phi'(x)|$$

where $\|\mu\|$ is the total variation of μ. This means that the Stieltjes convolution is a distribution of order one or zero. One easily shows that $\mu\Delta\lambda$ belongs to $\mathcal{O}'_{-1}$ so that its Stieltjes transform exists. An important property is

THEOREM A.2. : If μ and λ are two complex measures on $\mathbb{R}$ then

(A.2.5) $$S(\mu\Delta\lambda;z) = S(\mu;z)S(\lambda;z) .$$

Proof : Straightforward calculus gives

$$S(\mu\Delta\lambda;z) = \iint \left(\frac{1}{z-x} - \frac{1}{z-y}\right) d\mu(x)\, d\lambda(y)$$

$$= \iint \frac{1}{z-x}\, \frac{1}{z-y}\, d\mu(x)\, d\lambda(y) .$$

■

This property (A.2.5) justifies the word convolution (compare this with the convolution theorem for Laplace and Fourier transforms). The anti derivative of $\mu\Delta\lambda$ is denoted by $\mu\nabla\lambda$ and proves to be a measure with distribution function

(A.2.6) $$\mu\nabla\lambda(x) = \iint \frac{(x-t)U(x-t) - (x-y)U(x-y)}{y - t} d\mu(t)\, d\lambda(y)$$

where U is the Heaviside function. This is indeed a measure since the integrand in (A.2.6) is positive and bounded by 1. This anti derivative has a nice probabilistic interpretation. If X and Y are two independent random variables then we say that Z is uniformly distributed between X and Y if

$$P\{Z \leq t | X = x, Y = y\} = 1 \qquad t \geq x , t \geq y$$

$$= 0 \qquad t < x , t < y$$

$$= \frac{t-x}{y-x} \qquad x \leq t < y$$

$$= \frac{t-y}{x-y} \qquad y \leq t < x ,$$

which means that, given X = x and Y = y, the variable Z is uniformly distributed on the interval [min(x,y), max(x,y)].

THEOREM A.3. : Let X and Y be two independent random variables with distribution functions given by μ and λ, then the random variable Z, which is uniformly distributed between X and Y, has $\mu\nabla\lambda$ as its distribution function.

REFERENCES

[1] M. ABRAMOWITZ, I.A. STEGUN : "Handbook of Mathematical Functions", Dover Publications, New York, 1965.

[2] N.I. AKHIEZER : "The Classical Moment Problem" Oliver and Boyd, Edinburgh, 1965.

[3] N.I. AKHIEZER : *Orthogonal polynomials on several intervals*, Dokl. Akad. Nauk. CCCP 134 (1960), 9-12 (Russian); Soviet Math. Dokl. 1 (1961), 989-992.

[4] L. ARNOLD : *On the asymptotic distribution of the eigenvalues of random matrices*, J. Math. Anal. Appl. 20 (1967), 262-268.

[5] L. ARNOLD : *On Wigner's semicircle law for the eigenvalues of random matrices*, Z. Wahrsch. verw. Gebiete 19 (1971), 191-198.

[6] L. ARNOLD : *The possible asymptotic distributions of eigenvalues of random symmetric matrices*, Trans. 6th Prague Conference on Information Theory - Statistical Decision Functions - Random Processes (1971), 51-62.

[7] L. ARNOLD : *Deterministic version of Wigner's semicircle law for the distribution of matrix eigenvalues*, Linear Algebra Appl. 13 (1976), 185-199.

[8] R. ASKEY, M. ISMAIL : *Recurrence relations, continued fractions and orthogonal polynomials*, Mem. Amer. Math. Soc. 300 (1984), Amer. Math. Soc., Providence, R.I.

[9] R. ASKEY, J. WILSON : *Some basic hypergeometric orthogonal polynomials that generalize Jacobi polynomials*, Mem. Amer. Math. Soc. 319 (1985), Amer. Math. Soc., Providence, R.I.

[10] F.V. ATKINSON : "Discrete and Continuous Boundary Value Problems", Academic Press, New York, 1964.

[11] G.A. BAKER, D. BESSIS, P. MOUSSA : *A family of almost periodic Schrödinger operators*, Physica 124A (1984), 61-78.

[12] M.F. BARNSLEY, J.S. GERONIMO, A.N. HARRINGTON : *Orthogonal polynomials associated with invariant measures on Julia sets*, Bull. Amer. Math. Soc. 7 (1982),

381-384.

[13] M.F. BARNSLEY, J.S. GERONIMO, A.N. HARRINGTON : *On the invariant sets of a family of quadratic mappings,* Comm. Math. Phys. 88 (1983), 479-501.

[14] M.F. BARNSLEY, J.S. GERONIMO, A.N. HARRINGTON : *Geometry, electrostatic measure and orthogonal polynomials on Julia sets for polynomials,* Ergodic Theory Dynamical Systems 3 (1983), 509-520.

[15] M.F. BARNSLEY, J.S. GERONIMO, A.N. HARRINGTON : *Infinite-dimensional Jacobi matrices associated with Julia sets,* Proc. Amer. Math. Soc. 88 (1983), 625-630.

[16] W. BARRETT : *An asymptotic formula relating to orthogonal polynomials,* J. London Math. Soc. (2) 6 (1973), 701-704.

[17] P. BARRUCAND, D. DICKINSON : *On cubic transformations of orthogonal polynomials,* Proc. Amer. Math. Soc. 17 (1966), 810-814.

[18] W.C. BAULDRY : *Orthogonal polynomials associated with exponential weights,* Doctoral dissertation, Ohio State University (1985).

[19] W. BAULDRY, A. MATE, P. NEVAI : *Asymptotic expansions of recurrence coefficients of asymmetric Freud polynomials,* in "Approximation V".

[20] W.C. BAULDRY, A. MATE, P. NEVAI : *Asymptotics for solutions of systems of smooth recurrence equations,* manuscript.

[21] J. BELLISARD, D. BESSIS, P. MOUSSA : *Chaotic states of almost periodic Schrödinger operators,* Phys. Rev. Lett. 49 (1982), 701-704.

[22] S.N. BERNSTEIN : *Sur les polynômes orthogonaux relatifs à un segment fini, I, II,* J. Math. Pures Appl. 9 (1930), 127-177; 10 (1931), 216-286.

[23] D. BESSIS : *A new method in the combinatorics of the topological expansion,* Comm. Math. Phys. 69 (1979), 147-163.

[24] D. BESSIS, J.S. GERONIMO, P. MOUSSA : *Function weighted measures and orthogonal polynomials on Julia sets,* manuscript.

[25] D. BESSIS, C. ITZYKSON, J.B. ZUBER : *Quantum field techniques in graphical enumeration,* Adv. in Appl. Math. 1 (1980), 109-157.

[26] D. BESSIS, M.L. MEHTA, P. MOUSSA : *Orthogonal polynomials on a family of Cantor sets and the problem of iterations of quadratic mappings,* Lett. Math. Phys. 6 (1982), 123-140.

[27] D. BESSIS, P. MOUSSA : *Orthogonality properties of iterated polynomial mappings,* Comm. Math. Phys. 88 (1983), 503-529.

[28] P. BILLINGSLEY : "Convergence of Probability Measures", John Wiley, New York, 1968.

[28'] N.H. BINGHAM, C.M. GOLDIE, J.L. TEUGELS : "Regular variation", Encycl. of Math., Cambridge University Press, 1986.

[29] O. BLUMENTHAL : *Uber die Entwicklung einer willkürlichen Funktion nach den Nennern des Kettenbruches für* $\int_{-\infty}^{0}[\phi(\xi)/(z-\xi)]d\xi$, Inaugural Dissertation, Göttingen, 1898.

[30] R.P. BOAS : "Entire Functions", Academic Press, New York, 1959.

[31] H. BREMERMANN : "Distributions, Complex Variables and Fourier Transforms", Addison-Wesley, Reading, Massachusetts, 1965.

[32] H. BROLIN : *Invariant sets under iteration of rational functions,* Ark. Mat. 6 (1965), 103-144.

[33] H. CALLAERT : *On the rate of convergence in birth-and-death processes,* Bull. Soc. Math. Belg. 26 (1974), 173-184.

[34] K.M. CASE : *Orthogonal polynomials from the viewpoint of scattering theory,* J. Math. Phys. 15 (1974), 2166-2174.

[35] K.M. CASE, M. KAC : *A discrete version of the inverse scattering problem,* J. Math. Phys. 14 (1973), 594-603.

[36] C.V.L. CHARLIER : *Uber die Darstellung willkurlicher Funktionen,* Ark. Mat. Astr. och Fysic 2 (1905-1906), 20.

[37] P.L. CHEBYSHEV : *Sur le développement des fonctions à une seule variable,* Bull. Phys.-Math. Acad. Imp. Sci. St. Petersburg I (1859), 193-200.

[38] P.L. CHEBYSHEV : *Sur les valeurs limites des intégrales,* J. Math. Pures Appl. 19 (1874), 157-160.

[38'] P.L. CHEBYSHEV : *Sur les fractions continues,* Usp. Zap. Imper. Akad. Nauk. 3 (1855), 636-664 (Russian); J. Math. pures Appl. (II), 3 (1858), 289-323.

[39] T.S. CHIHARA : "An Introduction to Orthogonal Polynomials", Gordon and Breach, New York, 1978.

[40] T.S. CHIHARA : *Chain sequences and orthogonal polynomials,* Trans. Amer. Math.

104 (1962), 1-16.

[41] T.S. CHIHARA : *On kernel polynomials and related systems,* Boll. Un. Mat. Ital. (3) 19 (1964), 451-459.

[42] T.S. CHIHARA : *On recursively defined orthogonal polynomials,* Proc. Amer. Math. Soc. 16 (1965), 702-710.

[43] T.S. CHIHARA : *Orthogonal polynomials whose zeros are dense in intervals,* J. Math. Anal. Appl. 24 (1968), 362-371.

[44] T.S. CHIHARA : *Spectral properties of orthogonal polynomials on unbounded sets,* Trans. Amer. Math. Soc. 270 (1982), 623-639.

[45] T.S. CHIHARA, M.E.H. ISMAIL : *Orthogonal polynomials suggested by a queueing model,* Adv. in Appl. Math. 3 (1982), 441-462.

[46] E.B. CHRISTOFFEL : *Uber die Gaussische Quadratur und eine Verallgemeinung derselben,* J. Reine Angew. Math. 55 (1858), 61-82.

[47] G. DARBOUX : *Mémoires sur l'approximation des fonctions de très grands nombres,* J. Math. (3) 4 (1878), 5-56; 377-416.

[48] J.S. DEHESA : *Lanczos method of tridiagonalization, Jacobi matrices and physics,* J. Comput. Appl. Math. 7 (1981), 249-259.

[49] F. DELYON, B. SIMON, B. SOUILLARD : *From power pure point to continuous spectrum in disordered systems,* Ann. Inst. H. Poincaré, Phys. Théor. 42 (1985), 283-309.

[50] J. DOMBROWSKI, P. NEVAI : *Orthogonal polynomials, measures and recurrence relations,* SIAM J. Math. Anal. 17 (1986), 752-759.

[51] A. ERDELYI : *Asymptotic forms for Laguerre polynomials,* J. Indian Math. Soc. 24 (1960), 235-250.

[52] P. ERDOS : *On the distribution of the roots of orthogonal polynomials,* Proceedings of the Conference on Constructive Theory of Functions (G. Alexits, S.B. Stechkin, ed.), Akadémiai Kiado, Budapest, 1972, 145-150.

[53] P. ERDOS, G. FREUD : *On orthogonal polynomials with regularly distributed zeros,* Proc. London Math. Soc. (3) 29 (1974), 521-537.

[54] P. ERDOS, P. TURAN : *On interpolation, III,* Ann. of Math. 41 (1940), 510-555.

[55] G. FANO, G. GALLAVOTTI : *Dense sums,* Ann. Inst. H. Poincaré (A) 17 (1972),

195-219.

[56] G. FANO, G. LOUPIAS : *On the thermodynamical limit of the BCS state,* Comm. Math. Phys. 20 (1971), 143-166.

[57] P. FATOU : *Sur les équations fonctionelles, I, II, III,* Bull. Soc. Math. France 47 (1919), 161-271; 48 (1920), 33-94; 208-314.

[58] J. FAVARD : *Sur les polynômes de Tchebicheff,* C.R. Acad. Sci. Paris, 200 (1935), 2052-2053.

[59] E. FELDHEIM : *On a system of orthogonal polynomials associated with a distribution of Stieltjes type,* C.R. Acad. Sci. URSS 31 (1941), 528-533.

[60] M.E. FISHER : *The free energy of a macroscopic system,* Arch. Rational Mech. Anal. 17 (1964), 377-410.

[61] G. FREUD : "Orthogonal Polynomials", Pergamon Press, Oxford, 1971.

[62] G. FREUD : *Uber die Asymptotik Orthogonaler Polynome* Publ. Inst. Math. (Beograd) 11 (1957), 19-32.

[63] G. FREUD : *On the greatest zero of an orthogonal polynomial, I, II,* Acta Sci. Math. (Szeged) 34 (1973), 91-97; 36 (1974), 49-54.

[64] G. FREUD : *On estimations of the greatest zeroes of orthogonal polynomials,* Acta Math. Acad. Sci. Hungar. 25 (1974), 99-107.

[65] G. FREUD : *On the coefficients in the recursion formula of orthogonal polynomials,* Proc. Royal Irish Acad. 76 (1976), 1-6.

[66] G. FREUD : *On the zeros of orthogonal polynomials with respect to measures with noncompact support,* Anal. Numér. Théor. Approx. 6 (1977), 125-131.

[67] G. FREUD : *On the greatest zero of an orthogonal polynomial,* J. Approx. Theory 46 (1986), 16-24.

[68] K.F. GAUSS : *Methodus nova integralium valores per approximationem inveniendi,* Comm. Soc. Sci. Gottingensis Rec. III (1816).

[69] J.S. GERONIMO : *A relation between the coefficients in the recurrence formula and the spectral function for orthogonal polynomials,* Trans. Amer. Math. Soc. 260 (1980), 65-82.

[70] J.S. GERONIMO : *Scattering theory and matrix orthogonal polynomials on the real line,* Circuits Systems Signal Proces. 1 (1982), 471-495.

[71] J.S. GERONIMO : *On the spectra of infinite dimensional Jacobi matrices,* J. Approx. Theory.

[72] J.S. GERONIMO, K.M. CASE : *Scattering theory and polynomials orthogonal on the real line,* Trans. Amer. Math. Soc. 258 (1980), 467-494.

[73] J.S. GERONIMO, P.G. NEVAI : *Necessary and sufficient conditions relating the coefficients in the recurrence formula to the spectral function for orthogonal polynomials,* SIAM J. Math. Anal. 14 (1983), 622-637.

[74] J.S. GERONIMO, W. VAN ASSCHE : *Orthogonal polynomials with asymptotically periodic recurrence coefficients,* J. Approx. Theory 46 (1986), 251-283.

[75] Ya.L. GERONIMUS : "Polynomials Orthogonal on a Circle and Interval", Pergamon Press, Oxford, 1960.

[76] Ya.L. GERONIMUS : *On mean and uniform approximation,* Dokl. Akad. Nauk. CCCP 88 (1953), 597-599 (in Russian).

[77] Ya.L. GERONIMUS : *Certain limiting properties of orthogonal polynomials,* Vestnik Kharkov Gos. Univ. 32 (1966), 40-50 (in Russian).

[78] Ya.L. GERONIMUS : *On the character of the solutions of the moment problem in the case of a limit-periodic associated fraction,* Izv. Akad. Nauk. CCCP 5 (1941), 203-210 (in Russian).

[79] Ya.L. GERONIMUS : *On some finite difference equations and corresponding systems of orthogonal polynomials,* Zap. Mat. Otd. Fiz.-Mat. Fak. i. Kharkov Mat. Obsc. (4) 25 (1957), 87-100 (in Russian).

[80] Ya.L. GERONIMUS : *Orthogonal polynomials (appendix to the Russian edition of Szegö's book),* Amer. Math. Soc. Transl. (2) 108 (1977), 37-130.

[81] A.A. GONCHAR, E.A. RAKHMANOV : *Equilibrium measure and distribution of zeros of extremal polynomials,* Mat. Sb. 125 (167) (1984), 117-127 (Russian); Math. USSR Sb. 53 (1986), 119-130.

[82] U. GRENANDER, G. SZEGO : "Toeplitz Forms and their Applications", Univ. California Press, Berkeley, 1958.

[83] G.H. HARDY, J.E. LITTLEWOOD, G. POLYA : "Inequalities", Cambridge Univ. Press, London, 1952.

[84] C. HERMITE : *Sur un nouveau développement en série de fonctions,* C.R. Acad. Sci. Paris 58 (1864), 93-100; 266-273.

[85] E. HILLE : "Analytic Function Theory (vol. II)", Ginn and Company, New York, 1962.

[86] D.R. HOFSTADTER : *Energy levels and wave functions of Bloch electrons in rational and irrational magnetic fields,* Phys. Rev. B 14 (1976), 2239-2249.

[87] E. INONU : *Orthogonality of a set of polynomials encountered in neutron-transport and radiative transfer theory,* J. Math. Phys. 11 (1970), 568-577.

[88] M.E.H. ISMAIL : *On sieved orthogonal polynomials III : orthogonality on several intervals,* Trans. Amer. Math. Soc. 294 (1986), 89-111.

[89] C.G.J. JACOBI : *Uber Gauss' neue Methode, die Werthe der Integrale näherungsweise zu finden,* J. Reine Angew. Math. 1 (1826), 301-308.

[90] C.G.J. JACOBI : *Untersuchungen über die Differentialgleichung der hypergeometrische Reihe,* J. Reine Angew. Math. 56 (1859), 149-165.

[91] A.T. JAMES : *Distributions of matrix variates and latent roots derived from normal samples,* Ann. Math. Statist. 35 (1964), 475-501.

[92] G. JULIA : *Mémoire sur l'itération des fonctions rationelles,* J. Math. Pures Appl. (7) 1 (1918), 47-245.

[93] M. KAC, P. VAN MOERBEKE : *On some periodic Toda lattices,* Proc. Nat. Acad. Sci. USA 72 (1975), 1627-1629.

[94] S. KARLIN, J. McGREGOR : *Representation of a class of stochastic processes,* Proc. Nat. Acad. Sci. USA 41 (1955), 387-391.

[95] S. KARLIN, J. McGREGOR : *The differential equations of birth and death processes and the Stieltjes moment problem,* Trans. Amer. Math. Soc. 85 (1957), 489-546.

[96] S. KARLIN, J. McGREGOR : *The classification of birth and death processes,* Trans. Amer. Math. Soc. 86 (1957), 366-400.

[97] S. KARLIN, J. McGREGOR : *Many server queueing processes with Poisson input and exponential service times,* Pacific J. Math. 8 (1958), 87-118.

[98] S. KARLIN, J. McGREGOR : *Linear growth birth and death processes,* J. Math. Mech. 7 (1958), 643-662.

[99] S. KARLIN, J. McGREGOR : *Random walks,* Illinois J. Math. 3 (1959), 66-81.

[100] T. KATO : "Perturbation Theory for Linear Operators", Springer-Verlag, New York, 1966.

[101] O.D. KELLOGG : "Foundations of Potential Theory", Springer-Verlag, Berlin, 1967.

[102] A.N. KOLMOGOROV : *Stationary sequences in Hilbert spaces,* Izv. Akad. Nauk. CCCP 5 (1941), 3-14 (in Russian).

[103] M. KREIN : *Concerning a special class of entire and meromorphic functions,* in "Some questions in the theory of moments, vol. 2", Transl. of Math. Monographs, 1962.

[104] M.G. KREIN : *On a generalization of the investigations of G. Szegö, V.I. Smirnov and A.N. Kolmogorov,* Dokl. Akad. Nauk. CCCP 46 (1945), 95-98 (in Russian).

[104'] J. LAGRANGE : *Mémoire sur l'utilité de la méthode de prendre le milieu entre les résultats de plusieurs observations,* Miscellanea Taurinensia V, 1770-1773.

[105] E.N. LAGUERRE : *Sur l'intégrale* $\int_x^\infty x^{-1} e^{-x} dx$, Bull. Soc. Math. France 7 (1879), 72-81.

[106] P.S. DE LAPLACE : *Traité de Mécanique Céleste,* (1799-1823).

[107] A.M. LEGENDRE : *Recherches sur l'attraction des sphéroides homogènes,* Mém. Math. Phys., présentés à l'Acad. Sciences 10 (1785).

[108] D.S. LUBINSKY, H.N. MHASKAR, E.B. SAFF : *Freud's conjecture for exponential weights,* Bull. Amer. Math. Soc. 15 (1986), 217-221.

[109] D.S. LUBINSKY, H.N. MHASKAR, E.B. SAFF : *A proof of Freud's conjecture for exponential weights,* manuscript.

[110] D.S. LUBINSKY, E.B. SAFF : *Strong asymptotics for extremal errors and extremal polynomials associated with weights on* $(-\infty,\infty)$, manuscript.

[111] M. MAEJIMA, W. VAN ASSCHE : *Probabilistic proofs of asymptotic formulas for some classical polynomials,* Math. Proc. Cambridge Philos. Soc. 97 (1985), 499-510.

[112] A.P. MAGNUS : *A proof of Freud's conjecture about the orthogonal polynomials related to* $|x|^\rho \exp(-x^{2m})$ *for integer* m, in "Polynômes orthogonaux et applications", Lecture Notes in Mathematics 1171, Springer-Verlag, Berlin,

1985, 362-372.

[113] A.P. MAGNUS : *On Freud's equations for exponential weights,* J. Approx. Theory 46 (1986), 65-99.

[114] D.P. MAKI : *On determining regular behavior from the recurrence formula for orthogonal polynomials,* Pacific J. Math. 91 (1980), 173-178.

[115] V.A. MARCHENKO, L.A. PASTUR : *Distribution of eigenvalues for some sets of random matrices,* Mat. Sb. 72 (114)(1967),... - ... (Russian); Math. USSR Sb. 1 (1967), 457-483.

[116] A. MATE, P. NEVAI : *Remarks on E.A. Rahmanov's paper "On the asymptotics of the ratio of orthogonal polynomials",* J. Approx. Theory 36 (1982), 64-72.

[117] A. MATE, P. NEVAI : *Orthogonal polynomials and absolutely continuous measures,* in "Approximation IV", (ed. C.K. Chui et al.) Academic Press, New York 1983, 611-617.

[118] A. MATE, P. NEVAI : *Sublinear perturbations of the differential equation* $y^{(n)} = 0$ *and of the analogous difference equation,* J. Differential Equations, 53 (1984), 234-257.

[119] A. MATE, P. NEVAI : *Asymptotics for solutions of smooth recurrence equations,* Proc. Amer. Math. Soc. 93 (1985), 423-429.

[120] A. MATE, P. NEVAI, V. TOTIK : *Mean Cesàro summability of orthogonal polynomials,* "Constructive Theory of functions '84", Proceedings, Sofia 1984, 588-599.

[121] A. MATE, P. NEVAI, V. TOTIK : *What is beyond Szegö's theory of orthogonal polynomials ?* in "Rational approximation and interpolation", Lecture Notes in Mathematics 1105, Springer-Verlag, Berlin, 1984, 502-510.

[122] A. MATE, P. NEVAI, V. TOTIK : *Asymptotics for the ratio of leading coefficients of orthonormal polynomials on the unit circle,* Constr. Approx. 1 (1985), 63-69.

[123] A. MATE, P. NEVAI, V. TOTIK : *Asymptotics for orthogonal polynomials defined by a recurrence relation,* Constr. Approx. 1 (1985), 231-248.

[124] A. MATE, P. NEVAI, V. TOTIK : *Asymptotics for the greatest zeros of orthogonal polynomials,* SIAM J. Math. Anal. 17 (1986), 745-751.

[125] A. MATE, P. NEVAI, V. TOTIK : *Asymptotics for the zeros of orthogonal polynomials associated with infinite intervals,* J. London Math. Soc. (2) 23 (1986),

303-310.

[126] A. MATE, P. NEVAI, V. TOTIK : *Extensions of Szegö's theory of orthogonal polynomials, II,* Constr. Approx. 3 (1987), 51-72.

[127] A. MATE, P. NEVAI, V. TOTIK : *Strong and weak convergence of orthogonal polynomials,* Amer. J. Math.

[128] A. MATE, P. NEVAI, V. TOTIK : *Twisted difference operators and perturbed Chebyshev polynomials,* manuscript.

[129] A. MATE, P. NEVAI, T. ZASLAVSKY : *Asymptotic expansions of ratios of coefficients of orthogonal polynomials with exponential weights,* Trans. Amer. Math. Soc. 287 (1985), 495-505.

[130] M.L. MEHTA : "Random Matrices", Academic Press, New York, 1967.

[131] J. MEIXNER : *Orthogonale Polynomsysteme mit einem besonderen Gestalt der erzeugenden Funktion,* J. London Math. Soc. 9 (1934), 6-13.

[132] H.N. MHASKAR, E.B. SAFF : *Extremal problems for polynomials with Laguerre weights,* in "Approximation IV" (ed. C.K. Chui et al.), Academic Press, New York, 1983, 619-624.

[133] H.N. MHASKAR, E.B. SAFF : *Extremal problems for polynomials with exponential weights,* Trans. Amer. Math. Soc. 285 (1984), 203-234.

[134] H.N. MHASKAR, E.B. SAFF : *Weighted polynomials on finite and infinite intervals : a unified approach,* Bull. Amer. Math. Soc. 11 (1984), 351-354.

[135] E. MOECKLIN : *Asymptotische Entwicklungen der Laguerreschen Polynome,* Comment. Math. Helv. 7 (1934), 24-46.

[136] P. MOUSSA : *Un opérateur de Schrödinger presque périodique à spectre singulier associé aux itérations d'un polynôme,* Comptes Rendus de la RCP 25 du CNRS, 34 (1984), 43-66.

[137] P. MOUSSA : *Itérations des polynômes et propriétés d'orthogonalité,* Ann. Inst. H. Poincaré, Phys. Théor. 44 (1986), 315-325.

[138] P.G. NEVAI : *Orthogonal Polynomials,* Mem. Amer. Math. Soc. 213 (1979), Amer. Math. Soc., Providence, R.I.

[139] P.G. NEVAI : *Orthonormal polynomials with weight function* $|x|^{\alpha}e^{-|x|^{\beta}}$, I, Acta Math. Acad. Sci. Hungar. 24 (1973), 335-342 (in Russian).

[140] P.G. NEVAI : *Some properties of polynomials orthogonal with weight* $(1+x^{2k})^{\alpha}$ $e^{-|x|^{2k}}$ *and their applications in approximation theory,* Dokl. Akad. Nauk. CCCP 211 (1973), 784-786 (in Russian); Soviet Math. Dokl. 14 (1973), 1116-1119.

[141] P.G. NEVAI : *On orthogonal polynomials,* J. Approx. Theory 25 (1979), 34-37.

[142] P.G. NEVAI : *Distribution of zeros of orthogonal polynomials,* Trans. Amer. Math. Soc. 249 (1979), 341-361.

[143] P.G. NEVAI : *Orthogonal polynomials defined by a recurrence relation,* Trans. Amer. Math. Soc. 250 (1979), 369-384.

[144] P.G. NEVAI : *Eigenvalue distribution of Toeplitz matrices,* Proc. Amer. Math. Soc. 80 (1980), 247-253.

[145] P.G. NEVAI : *Orthogonal polynomials associated with* $\exp(-x^4)$, Canad. Math. Soc., Conference Proceedings 3 (1983), 263-285.

[146] P. NEVAI : *A new class of orthogonal polynomials,* Proc. Amer. Math. Soc. 91 (1984), 409-415.

[147] P. NEVAI : *Asymptotics for orthogonal polynomials associated with* $\exp(-x^4)$, SIAM J. Math. Anal. 15 (1984), 1177-1187.

[148] P.G. NEVAI : *Two of my favorite ways of obtaining asymptotics for orthogonal polynomials,* Funct. Anal. and Approx., Birkhaüser Verlag, Basel, 1984.

[149] P.G. NEVAI : *Orthogonal polynomials on infinite intervals,* Rend. Sem. Mat. Univ. Politec. Torino, july 1985, 215-235.

[150] P. NEVAI : *Extensions of Szegö's theory of orthogonal polynomials,* in "Polynômes orthogonaux et applications", Lecture Notes in Mathematics 1171, Springer-Verlag, Berlin, 1985, 230-238.

[151] P. NEVAI : *Géza Freud, orthogonal polynomials and Christoffel Functions. A case study,* J. Approx. Theory 48 (1986), 3-167.

[152] P.G. NEVAI, J.S. DEHESA : *On asymptotic average properties of zeros of orthogonal polynomials,* SIAM J. Math. Anal. 10 (1979), 1184-1192.

[153] O. PERRON : "Die Lehre von den Kettenbruchen", Chelsea Publishing Company, New York, 1950.

[154] V.V. PETROV : "Sums of Independent Random Variables", Springer-Verlag, Berlin, 1975.

[155] T.S. PITCHER, J.R. KINNEY : *Some connections between ergodic theory and the iteration of polynomials,* Ark. Mat. 8 (1968), 25-32.

[156] M. PLANCHEREL, W. ROTACH : *Sur les valeurs asymptotiques des polynômes d'Hermite,* Comment. Math. Helv. 1 (1929), 227-254.

[157] H. POINCARE : *Sur les équations linéaires aux différentielles ordinaires et aux différences finies,* Amer. J. Math. 7 (1885), 203-258.

[158] F. POLLACZEK : *Sur une généralisation des polynômes de Legendre,* C.R. Acad. Sci. Paris 228 (1949), 1363-1365.

[159] F. POLLACZEK : *Système de polynômes biorthogonaux qui généralisent les polynômes ultrasphériques,* C.R. Acad. Sci. Paris 228 (1949), 1998-2000.

[160] F. POLLACZEK : *Sur une généralisation des polynômes de Jacobi,* Mémoirs Sci. Math. 131 (1956), Gauthiers-Villars, Paris.

[161] E.A. RAKHMANOV : *On the asymptotics of the ratio of orthogonal polynomials,* Mat. Sb. 103 (145) (1977), 237-252; Math. USSR Sb. 32 (1977), 199-213.

[162] E.A. RAKHMANOV : *On the asymptotics of the ratio of orthogonal polynomials, II,* Math. Sb 118 (160)(1982), 104-117; Math. USSR Sb. 46 (1983), 105-117.

[163] E.A. RAKHMANOV : *On asymptotics properties of polynomials orthogonal on the real axis,* Mat. Sb. 119 (161) (1982), 163-203 (in Russian); Math. USSR Sb. 47 (1984), 155-193.

[164] A.W. ROBERTS, D.E. VARBERG : "Convex Functions", Academic Press, New York, 1973.

[165] W. RUDIN : "Principles of Mathematical Analysis", McGraw-Hill, Tokyo, 1964.

[166] W. RUDIN : "Functional Analysis", Tata McGraw-Hill, New Delhi, 1974.

[167] W. RUDIN : "Real and Complex Analysis", Tata McGraw-Hill, New Dehli, 1974.

[168] L. SARIO, M. NAKAI : "Classification Theory of Riemann Surfaces", Springer-Verlag, Berlin, 1970.

[169] E. SENETA : "Regularly varying functions", Lecture Notes in Mathematics 508, Springer-Verlag, Berlin, 1976.

[170] J.A. SHOHAT, J.D. TAMARKIN : "The Problem of Moments", Amer. Math. Soc., Providence, R.I., 1943.

[171] N.J. SONIN : *Recherches sur les fonctions cylindriques et le développement de fonctions continues en séries*, Math. Ann. 16 (1880), 1-80.

[172] W. STADJE : *Probabilistic proofs of some formulae for Bessel functions*, Indag. Math. 45 (1983), 343-359.

[173] T.J. STIELTJES : *Quelques recherches sur la théorie des quadratures dites mécaniques*, Ann Sci. Ecole. Norm. Sup. (3) 1 (1884), 409-426.

[174] T.J. STIELTJES : *Recherches sur les fractions continues*, Ann. Fac. Sci. Toulouse 8 (1894), J1-J122; 9 (1895), A1-A47.

[174'] M.H. STONE : *Linear transformations in Hilbert space and their applications to analysis*, Amer. Math. Soc. Colloq. Publ. 15, Providence, R.I., 1932.

[175] G. SZEGO : "Orthogonal Polynomials", Amer. Math. Soc. Colloq. Publ. 23, Providence, R.I., 4th edition 1975.

[176] G. SZEGO : *Hankel forms*, Mat. Term. Tud. Ert. 36 (1918), 497-538 (in Hungarian); Amer. Math. Soc. Transl. (2) 108 (1977), 1-36.

[177] G. SZEGO : *Beiträge zur Theorie der Toeplitzschen Formen, II*, Math. Z. 9 (1921), 167-190.

[178] W.C. TAYLOR : *A complete set of asymptotic formulas for the Whittaker function and the Laguerre polynomials*, J. Math. and Phys. (MIT) 18 (1939), 34-49.

[179] M. TSUJI : "Potential Theory in Modern Function Theory", Chelsea Publishing Company, New York, 1959.

[180] J.L. ULLMAN : *On the regular behaviour of orthogonal polynomials*, Proc. London Math. Soc. (3) 24 (1972), 119-148.

[181] J.L. ULLMAN : *Orthogonal polynomials for general measures*, in "Mathematical Structures", collection of papers devoted to Iliev, Sofia (1975), 493-496.

[182] J.L. ULLMAN : *A survey of exterior asymptotics for orthogonal polynomials associated with a finite interval and a study of the case of the general weight measures*, Proc. NATO Advanced Study Institute on Approx. Theory and Appl., Ser. C136, (1984), 467-478.

[183] J.L. ULLMAN : *Orthogonal polynomials for general measures, I*, in "Rational Approximation and interpolation", Lecture Notes in Mathematics 1105 (1984), Springer-Verlag, Berlin, 524-528.

[184] J.L. ULLMAN : *Orthogonal polynomials for general measures, II,* manuscript.

[185] J.L. ULLMAN : *Orthogonal polynomials associated with an infinite interval,* Michigan Math. J. 27 (1980), 353-363.

[186] Ch. J. DE LA VALLEE POUSSIN : "Le Potentiel Logarithmique", Gauthiers-Villars, Paris, 1949.

[187] W. VAN ASSCHE : *Weighted zero distribution for polynomials orthogonal on an infinite interval,* SIAM J. Math. Anal. 16 (1985), 1317-1334.

[188] W. VAN ASSCHE : *Some results on the asymptotic distribution of the zeros of orthogonal polynomials,* J. Comput. Appl. Math. 12 & 13 (1985), 615-623.

[189] W. VAN ASSCHE : *Asymptotic properties of orthogonal polynomials from their recurrence relation, I,* J. Approx. Theory 44 (1985), 258-276.

[190] W. VAN ASSCHE : *Invariant zero behaviour for orthogonal polynomials on compact sets of the real line,* Bull. Soc. Math. Belg. (B) (1986),

[191] W. VAN ASSCHE : *Asymptotic properties of orthogonal polynomials from their recurrence coefficients II,* J. Approx. Theory.

[192] W. VAN ASSCHE : *Eigenvalues of Toeplitz matrices associated with orthogonal polynomials,* J. Approx. Theory.

[193] W. VAN ASSCHE, G. FANO, F. ORTOLANI : *Asymptotic behaviour of the coefficients of some sequences of polynomials,* SIAM J. Math. Anal.

[194] W. VAN ASSCHE, J.S. GERONIMO : *Asymptotics for orthogonal polynomials on and off the essential spectrum,* J. Approx. Theory.

[195] W. VAN ASSCHE, J.S. GERONIMO : *Asymptotics for orthogonal polynomials with regularly varying recurrence coefficients,* Rocky Mountain J. Math.

[196] W. VAN ASSCHE, J.L. TEUGELS : *Second order asymptotic behaviour of the zeros of orthogonal polynomials,* Rev. Roumaine Math. Pures Appl. 32 (1987), 15-26.

[197] E.A. VAN DOORN : *The transient state probabilities for a queueing model where potential customers are discouraged by queue length,* J. Appl. Prob. 18 (1981), 499-506.

[198] P. VAN MOERBEKE : *The spectrum of periodic Jacobi matrices,* Invent. Math. 37 (1976), 45-81.

[199] H.S. WALL : "Analytic Theory of Continued Fractions", Chelsea Publishing Com-

pany, New York, 1973.

[200] J.C. WHEELER : *Modified moments and continued fraction coefficients for the diatomic linear chain*, J. Chem. Phys. 80 (1984), 472-476.

[201] H. WIDOM : *Polynomials associated with measures in the complex plane*, J. Math. Mech. 16 (1967), 997-1013.

[202] E.P. WIGNER : *Characteristic vectors of bordered matrices with infinite dimensions*, Ann. of Math. 62 (1955), 548-564.

[203] E.P. WIGNER : *Random matrices in physics*, Siam Review 9 (1967), 1-23.

[204] A. WINTNER : "Spektraltheorie der Unendlichen Matrizen", Hirzel, Leipzig, 1929.

[205] W. WOESS : *Random walks and periodic continued fractions*, Adv. Appl. Prob. 17 (1985), 67-84.

SUBJECT INDEX

LECTURE NOTES IN MATHEMATICS
Edited by A. Dold and B. Eckmann

Some general remarks on the publication of monographs and seminars

In what follows all references to monographs, are applicable also to multiauthorship volumes such as seminar notes.

1. Lecture Notes aim to report new developments - quickly, informally, and at a high level. Monograph manuscripts should be reasonably self-contained and rounded off. Thus they may, and often will, present not only results of the author but also related work by other people. Furthermore, the manuscripts should provide sufficient motivation, examples and applications. This clearly distinguishes Lecture Notes manuscripts from journal articles which normally are very concise. Articles intended for a journal but too long to be accepted by most journals, usually do not have this "lecture notes" character. For similar reasons it is unusual for Ph.D. theses to be accepted for the Lecture Notes series.

   Experience has shown that English language manuscripts achieve a much wider distribution.

2. Manuscripts or plans for Lecture Notes volumes should be submitted either to one of the series editors or to Springer-Verlag, Heidelberg. These proposals are then refereed. A final decision concerning publication can only be made on the basis of the complete manuscripts, but a preliminary decision can usually be based on partial information: a fairly detailed outline describing the planned contents of each chapter, and an indication of the estimated length, a bibliography, and one or two sample chapters - or a first draft of the manuscript. The editors will try to make the preliminary decision as definite as they can on the basis of the available information.

3. Lecture Notes are printed by photo-offset from typed copy delivered in camera-ready form by the authors. Springer-Verlag provides technical instructions for the preparation of manuscripts, and will also, on request, supply special staionery on which the prescribed typing area is outlined. Careful preparation of the manuscripts will help keep production time short and ensure satisfactory appearance of the finished book. Running titles are not required; if however they are considered necessary, they should be uniform in appearance. We generally advise authors not to start having their final manuscripts specially tpyed beforehand. For professionally typed manuscripts, prepared on the special stationery according to our instructions, Springer-Verlag will, if necessary, contribute towards the typing costs at a fixed rate.

   The actual production of a Lecture Notes volume takes 6-8 weeks.

.../...

4. Final manuscripts should contain at least 100 pages of mathematical text and should include

   - a table of contents
   - an informative introduction, perhaps with some historical remarks. It should be accessible to a reader not particularly familiar with the topic treated.
   - subject index; this is almost always genuinely helpful for the reader.

5. Authors receive a total of 50 free copies of their volume, but no royalties. They are entitled to purchase further copies of their book for their personal use at a discount of 33 1/3 %, other Springer mathematics books at a discount of 20 % directly from Springer-Verlag.

   Commitment to publish is made by letter of intent rather than by signing a formal contract. Springer-Verlag secures the copyright for each volume.

Vol. 1117: D.J. Aldous, J.A. Ibragimov, J. Jacod, Ecole d'Été de Probabilités de Saint-Flour XIII – 1983. Édité par P.L. Hennequin. X, 409 pages. 1985.

Vol. 1118: Grossissements de filtrations: exemples et applications. Séminaire, 1982/83. Edité par Th. Jeulin et M. Yor. V, 315 pages. 1985.

Vol. 1119: Recent Mathematical Methods in Dynamic Programming. Proceedings, 1984. Edited by I. Capuzzo Dolcetta, W.H. Fleming and T. Zolezzi. VI, 202 pages. 1985.

Vol. 1120: K. Jarosz, Perturbations of Banach Algebras. V, 118 pages. 1985.

Vol. 1121: Singularities and Constructive Methods for Their Treatment. Proceedings, 1983. Edited by P. Grisvard, W. Wendland and J.R. Whiteman. IX, 346 pages. 1985.

Vol. 1122: Number Theory. Proceedings, 1984. Edited by K. Alladi. VII, 217 pages. 1985.

Vol. 1123: Séminaire de Probabilités XIX 1983/84. Proceedings. Edité par J. Azéma et M. Yor. IV, 504 pages. 1985.

Vol. 1124: Algebraic Geometry, Sitges (Barcelona) 1983. Proceedings. Edited by E. Casas-Alvero, G.E. Welters and S. Xambó-Descamps. XI, 416 pages. 1985.

Vol. 1125: Dynamical Systems and Bifurcations. Proceedings, 1984. Edited by B.L.J. Braaksma, H.W. Broer and F. Takens. V, 129 pages. 1985.

Vol. 1126: Algebraic and Geometric Topology. Proceedings, 1983. Edited by A. Ranicki, N. Levitt and F. Quinn. V, 423 pages. 1985.

Vol. 1127: Numerical Methods in Fluid Dynamics. Seminar. Edited by F. Brezzi, VII, 333 pages. 1985.

Vol. 1128: J. Elschner, Singular Ordinary Differential Operators and Pseudodifferential Equations. 200 pages. 1985.

Vol. 1129: Numerical Analysis, Lancaster 1984. Proceedings. Edited by P.R. Turner. XIV, 179 pages. 1985.

Vol. 1130: Methods in Mathematical Logic. Proceedings, 1983. Edited by C.A. Di Prisco. VII, 407 pages. 1985.

Vol. 1131: K. Sundaresan, S. Swaminathan, Geometry and Nonlinear Analysis in Banach Spaces. III, 116 pages. 1985.

Vol. 1132: Operator Algebras and their Connections with Topology and Ergodic Theory. Proceedings, 1983. Edited by H. Araki, C.C. Moore, Ş. Strătilă and C. Voiculescu. VI, 594 pages. 1985.

Vol. 1133: K.C. Kiwiel, Methods of Descent for Nondifferentiable Optimization. VI, 362 pages. 1985.

Vol. 1134: G.P. Galdi, S. Rionero, Weighted Energy Methods in Fluid Dynamics and Elasticity. VII, 126 pages. 1985.

Vol. 1135: Number Theory, New York 1983–84. Seminar. Edited by D.V. Chudnovsky, G.V. Chudnovsky, H. Cohn and M.B. Nathanson. V, 283 pages. 1985.

Vol. 1136: Quantum Probability and Applications II. Proceedings, 1984. Edited by L. Accardi and W. von Waldenfels. VI, 534 pages. 1985.

Vol. 1137: Xiao G., Surfaces fibrées en courbes de genre deux. IX, 103 pages. 1985.

Vol. 1138: A. Ocneanu, Actions of Discrete Amenable Groups on von Neumann Algebras. V, 115 pages. 1985.

Vol. 1139: Differential Geometric Methods in Mathematical Physics. Proceedings, 1983. Edited by H. D. Doebner and J. D. Hennig. VI, 337 pages. 1985.

Vol. 1140: S. Donkin, Rational Representations of Algebraic Groups. VII, 254 pages. 1985.

Vol. 1141: Recursion Theory Week. Proceedings, 1984. Edited by H.-D. Ebbinghaus, G.H. Müller and G.E. Sacks. IX, 418 pages. 1985.

Vol. 1142: Orders and their Applications. Proceedings, 1984. Edited by I. Reiner and K. W. Roggenkamp. X, 306 pages. 1985.

Vol. 1143: A. Krieg, Modular Forms on Half-Spaces of Quaternions. XIII, 203 pages. 1985.

Vol. 1144: Knot Theory and Manifolds. Proceedings, 1983. Edited by D. Rolfsen. V, 163 pages. 1985.

Vol. 1145: G. Winkler, Choquet Order and Simplices. VI, 143 pages. 1985.

Vol. 1146: Séminaire d'Algèbre Paul Dubreil et Marie-Paule Malliavin. Proceedings, 1983–1984. Edité par M.-P. Malliavin. IV, 420 pages. 1985.

Vol. 1147: M. Wschebor, Surfaces Aléatoires. VII, 111 pages. 1985.

Vol. 1148: Mark A. Kon, Probability Distributions in Quantum Statistical Mechanics. V, 121 pages. 1985.

Vol. 1149: Universal Algebra and Lattice Theory. Proceedings, 1984. Edited by S. D. Comer. VI, 282 pages. 1985.

Vol. 1150: B. Kawohl, Rearrangements and Convexity of Level Sets in PDE. V, 136 pages. 1985.

Vol 1151: Ordinary and Partial Differential Equations. Proceedings, 1984. Edited by B.D. Sleeman and R.J. Jarvis. XIV, 357 pages. 1985.

Vol. 1152: H. Widom, Asymptotic Expansions for Pseudodifferential Operators on Bounded Domains. V, 150 pages. 1985.

Vol. 1153: Probability in Banach Spaces V. Proceedings, 1984. Edited by A. Beck, R. Dudley, M. Hahn, J. Kuelbs and M. Marcus. VI, 457 pages. 1985.

Vol. 1154: D.S. Naidu, A.K. Rao, Singular Pertubation Analysis of Discrete Control Systems. IX, 195 pages. 1985.

Vol. 1155: Stability Problems for Stochastic Models. Proceedings, 1984. Edited by V.V. Kalashnikov and V.M. Zolotarev. VI, 447 pages. 1985.

Vol. 1156: Global Differential Geometry and Global Analysis 1984. Proceedings, 1984. Edited by D. Ferus, R.B. Gardner, S. Helgason and U. Simon. V, 339 pages. 1985.

Vol. 1157: H. Levine, Classifying Immersions into $\mathbb{R}^4$ over Stable Maps of 3-Manifolds into $\mathbb{R}^2$. V, 163 pages. 1985.

Vol. 1158: Stochastic Processes – Mathematics and Physics. Proceedings, 1984. Edited by S. Albeverio, Ph. Blanchard and L. Streit. VI, 230 pages. 1986.

Vol. 1159: Schrödinger Operators, Como 1984. Seminar. Edited by S. Graffi. VIII, 272 pages. 1986.

Vol. 1160: J.-C. van der Meer, The Hamiltonian Hopf Bifurcation. VI, 115 pages. 1985.

Vol. 1161: Harmonic Mappings and Minimal Immersions, Montecatini 1984. Seminar. Edited by E. Giusti. VII, 285 pages. 1985.

Vol. 1162: S.J.L. van Eijndhoven, J. de Graaf, Trajectory Spaces, Generalized Functions and Unbounded Operators. IV, 272 pages. 1985.

Vol. 1163: Iteration Theory and its Functional Equations. Proceedings, 1984. Edited by R. Liedl, L. Reich and Gy. Targonski. VIII, 231 pages. 1985.

Vol. 1164: M. Meschiari, J.H. Rawnsley, S. Salamon, Geometry Seminar "Luigi Bianchi" II – 1984. Edited by E. Vesentini. VI, 224 pages. 1985.

Vol. 1165: Seminar on Deformations. Proceedings, 1982/84. Edited by J. Ławrynowicz. IX, 331 pages. 1985.

Vol. 1166: Banach Spaces. Proceedings, 1984. Edited by N. Kalton and E. Saab. VI, 199 pages. 1985.

Vol. 1167: Geometry and Topology. Proceedings, 1983–84. Edited by J. Alexander and J. Harer. VI, 292 pages. 1985.

Vol. 1168: S.S. Agaian, Hadamard Matrices and their Applications. III, 227 pages. 1985.

Vol. 1169: W.A. Light, E.W. Cheney, Approximation Theory in Tensor Product Spaces. VII, 157 pages. 1985.

Vol. 1170: B.S. Thomson, Real Functions. VII, 229 pages. 1985.

Vol. 1171: Polynômes Orthogonaux et Applications. Proceedings, 1984. Edité par C. Brezinski, A. Draux, A.P. Magnus, P. Maroni et A. Ronveaux. XXXVII, 584 pages. 1985.

Vol. 1172: Algebraic Topology, Göttingen 1984. Proceedings. Edited by L. Smith. VI, 209 pages. 1985.

Vol. 1173: H. Delfs, M. Knebusch, Locally Semialgebraic Spaces. XVI, 329 pages. 1985.

Vol. 1174: Categories in Continuum Physics, Buffalo 1982. Seminar. Edited by F.W. Lawvere and S.H. Schanuel. V, 126 pages. 1986.

Vol. 1175: K. Mathiak, Valuations of Skew Fields and Projective Hjelmslev Spaces. VII, 116pages. 1986.

Vol. 1176: R.R. Bruner, J.P. May, J.E. McClure, M. Steinberger, H_∞ Ring Spectra and their Applications. VII, 388 pages. 1986.

Vol. 1177: Representation Theory I. Finite Dimensional Algebras. Proceedings, 1984. Edited by V. Dlab, P. Gabriel and G. Michler. XV, 340 pages. 1986.

Vol. 1178: Representation Theory II. Groups and Orders. Proceedings, 1984. Edited by V. Dlab, P. Gabriel and G. Michler. XV, 370 pages. 1986.

Vol. 1179: Shi J.-Y. The Kazhdan-Lusztig Cells in Certain Affine Weyl Groups. X, 307 pages. 1986.

Vol. 1180: R. Carmona, H. Kesten, J.B. Walsh, École d'Été de Probabilités de Saint-Flour XIV – 1984. Édité par P.L. Hennequin. X, 438 pages. 1986.

Vol. 1181: Buildings and the Geometry of Diagrams, Como 1984. Seminar. Edited by L. Rosati. VII, 277 pages. 1986.

Vol. 1182: S. Shelah, Around Classification Theory of Models. VII, 279 pages. 1986.

Vol. 1183: Algebra, Algebraic Topology and their Interactions. Proceedings, 1983. Edited by J.-E. Roos. XI, 396 pages. 1986.

Vol. 1184: W. Arendt, A. Grabosch, G. Greiner, U. Groh, H.P. Lotz, U. Moustakas, R. Nagel, F. Neubrander, U. Schlotterbeck, One-parameter Semigroups of Positive Operators. Edited by R. Nagel. X, 460 pages. 1986.

Vol. 1185: Group Theory, Beijing 1984. Proceedings. Edited by Tuan H.F. V, 403 pages. 1986.

Vol. 1186: Lyapunov Exponents. Proceedings, 1984. Edited by L. Arnold and V. Wihstutz. VI, 374 pages. 1986.

Vol. 1187: Y. Diers, Categories of Boolean Sheaves of Simple Algebras. VI, 168 pages. 1986.

Vol. 1188: Fonctions de Plusieurs Variables Complexes V. Séminaire, 1979–85. Edité par François Norguet. VI, 306 pages. 1986.

Vol. 1189: J. Lukeš, J. Malý, L. Zajíček, Fine Topology Methods in Real Analysis and Potential Theory. X, 472 pages. 1986.

Vol. 1190: Optimization and Related Fields. Proceedings, 1984. Edited by R. Conti, E. De Giorgi and F. Giannessi. VIII, 419 pages. 1986.

Vol. 1191: A.R. Its, V.Yu. Novokshenov, The Isomonodromic Deformation Method in the Theory of Painlevé Equations. IV, 313 pages. 1986.

Vol. 1192: Equadiff 6. Proceedings, 1985. Edited by J. Vosmansky and M. Zlámal. XXIII, 404 pages. 1986.

Vol. 1193: Geometrical and Statistical Aspects of Probability in Banach Spaces. Proceedings, 1985. Edited by X. Fernique, B. Heinkel, M.B. Marcus and P.A. Meyer. IV, 128 pages. 1986.

Vol. 1194: Complex Analysis and Algebraic Geometry. Proceedings, 1985. Edited by H. Grauert. VI, 235 pages. 1986.

Vol.1195: J.M. Barbosa, A.G. Colares, Minimal Surfaces in $\mathbb{R}^3$. X, 124 pages. 1986.

Vol. 1196: E. Casas-Alvero, S. Xambó-Descamps, The Enumerative Theory of Conics after Halphen. IX, 130 pages. 1986.

Vol. 1197: Ring Theory. Proceedings, 1985. Edited by F.M.J. van Oystaeyen. V, 231 pages. 1986.

Vol. 1198: Séminaire d'Analyse, P. Lelong – P. Dolbeault – H. Skoda. Seminar 1983/84. X, 260 pages. 1986.

Vol. 1199: Analytic Theory of Continued Fractions II. Proceedings, 1985. Edited by W.J. Thron. VI, 299 pages. 1986.

Vol. 1200: V.D. Milman, G. Schechtman, Asymptotic Theory of Finite Dimensional Normed Spaces. With an Appendix by M. Gromov. VIII, 156 pages. 1986.

Vol. 1201: Curvature and Topology of Riemannian Manifolds. Proceedings, 1985. Edited by K. Shiohama, T. Sakai and T. Sunada. VII, 336 pages. 1986.

Vol. 1202: A. Dür, Möbius Functions, Incidence Algebras and Power Series Representations. XI, 134 pages. 1986.

Vol. 1203: Stochastic Processes and Their Applications. Proceedings, 1985. Edited by K. Itô and T. Hida. VI, 222 pages. 1986.

Vol. 1204: Séminaire de Probabilités XX, 1984/85. Proceedings. Edité par J. Azéma et M. Yor. V, 639 pages. 1986.

Vol. 1205: B.Z. Moroz, Analytic Arithmetic in Algebraic Number Fields. VII, 177 pages. 1986.

Vol. 1206: Probability and Analysis, Varenna (Como) 1985. Seminar. Edited by G. Letta and M. Pratelli. VIII, 280 pages. 1986.

Vol. 1207: P.H. Bérard, Spectral Geometry: Direct and Inverse Problems. With an Appendix by G. Besson. XIII, 272 pages. 1986.

Vol. 1208: S. Kaijser, J.W. Pelletier, Interpolation Functors and Duality. IV, 167 pages. 1986.

Vol. 1209: Differential Geometry, Peñíscola 1985. Proceedings. Edited by A.M. Naveira, A. Ferrández and F. Mascaró. VIII, 306 pages. 1986.

Vol. 1210: Probability Measures on Groups VIII. Proceedings, 1985. Edited by H. Heyer. X, 386 pages. 1986.

Vol. 1211: M.B. Sevryuk, Reversible Systems. V, 319 pages. 1986.

Vol. 1212: Stochastic Spatial Processes. Proceedings, 1984. Edited by P. Tautu. VIII, 311 pages. 1986.

Vol. 1213: L.G. Lewis, Jr., J.P. May, M. Steinberger, Equivariant Stable Homotopy Theory. IX, 538 pages. 1986.

Vol. 1214: Global Analysis – Studies and Applications II. Edited by Yu.G. Borisovich and Yu.E. Gliklikh. V, 275 pages. 1986.

Vol. 1215: Lectures in Probability and Statistics. Edited by G. del Pino and R. Rebolledo. V, 491 pages. 1986.

Vol. 1216: J. Kogan, Bifurcation of Extremals in Optimal Control. VIII, 106 pages. 1986.

Vol. 1217: Transformation Groups. Proceedings, 1985. Edited by S. Jackowski and K. Pawalowski. X, 396 pages. 1986.

Vol. 1218: Schrödinger Operators, Aarhus 1985. Seminar. Edited by E. Balslev. V, 222 pages. 1986.

Vol. 1219: R. Weissauer, Stabile Modulformen und Eisensteinreihen. III, 147 Seiten. 1986.

Vol. 1220: Séminaire d'Algèbre Paul Dubreil et Marie-Paule Malliavin. Proceedings, 1985. Edité par M.-P. Malliavin. IV, 200 pages. 1986.

Vol. 1221: Probability and Banach Spaces. Proceedings, 1985. Edited by J. Bastero and M. San Miguel. XI, 222 pages. 1986.

Vol. 1222: A. Katok, J.-M. Strelcyn, with the collaboration of F. Ledrappier and F. Przytycki, Invariant Manifolds, Entropy and Billiards; Smooth Maps with Singularities. VIII, 283 pages. 1986.

Vol. 1223: Differential Equations in Banach Spaces. Proceedings, 1985. Edited by A. Favini and E. Obrecht. VIII, 299 pages. 1986.

Vol. 1224: Nonlinear Diffusion Problems, Montecatini Terme 1985. Seminar. Edited by A. Fasano and M. Primicerio. VIII, 188 pages. 1986.

Vol. 1225: Inverse Problems, Montecatini Terme 1986. Seminar. Edited by G. Talenti. VIII, 204 pages. 1986.

Vol. 1226: A. Buium, Differential Function Fields and Moduli of Algebraic Varieties. IX, 146 pages. 1986.

Vol. 1227: H. Helson, The Spectral Theorem. VI, 104 pages. 1986.

Vol. 1228: Multigrid Methods II. Proceedings, 1985. Edited by W. Hackbusch and U. Trottenberg. VI, 336 pages. 1986.

Vol. 1229: O. Bratteli, Derivations, Dissipations and Group Actions on C*-algebras. IV, 277 pages. 1986.

Vol. 1230: Numerical Analysis. Proceedings, 1984. Edited by J.-P. Hennart. X, 234 pages. 1986.

Vol. 1231: E.-U. Gekeler, Drinfeld Modular Curves. XIV, 107 pages. 1986.